AN INTRODUCTION TO FIBER OPTICS

R. Allen Shotwell
Ivy Tech State College

Prentice Hall
Upper Saddle River, New Jersey   Columbus, Ohio

Library of Congress Cataloging-in-Publication Data

Shotwell, R. Allen.

    An introduction to fiber optics / by R. Allen Shotwell.

      p. cm.

    Includes index.

    ISBN 0-02-410172-9

    1. Fiber optics   I. Title.

TA1800. S53   1997

621.36′92—dc20   95–49050
                                                                 CIP

Cover photo: © Super Stock

Editor: Charles E. Stewart, Jr.

Production Editor: Julia Anderson Peters

Design Coordinator: Julia Zonneveld Van Hook

Text Design: Linda M. Robertson

Cover Design: Russ Maselli

Production Manager: Laura Messerly

Illustrations: Academy Artworks

Marketing Manager: Debbie Yarnell

This book was set in Century Schoolbook by Bi-Comp, Inc., and was printed and bound by Quebecor Printing/Book Press. The cover was printed by Phoenix Color Corp.

© 1997 by Prentice-Hall, Inc.

Simon & Schuster/A Viacom Company

Upper Saddle River, New Jersey 07458

Printed in the United States of America

10 9 8 7 6 5 4 3 2 1

ISBN: 0-02-410172-9

Prentice-Hall International (UK) Limited, *London*

Prentice-Hall of Australia Pty. Limited, *Sydney*

Prentic-Hall of Canada, Inc., *Toronto*

Prentice-Hall Hispanoamericana, S. A., *Mexico*

Prentice-Hall of India Private Limited, *New Delhi*

Prentice-Hall of Japan, Inc., *Tokyo*

Simon & Schuster Asia Pte. Ltd., *Singapore*

Editora Prentice-Hall do Brasil, Ltda., *Rio de Janeiro*

To my darling wife Christina . . .
Thank you for your patience and your love.

✳ PREFACE

The field of technical education has evolved to the point where, today, there are three distinct areas of study. At the theoretical level, students study technology's scientific aspects. In the design realm, engineering studies are necessary. From the practical, hands-on standpoint, with the need to acquire installation, repair, and operations knowledge and skills, students require technological and technician's courses. Although these three areas have only been relatively well defined for the past thirty or forty years, most basic/traditional areas of study have course materials that reflect all three. For instance, it is quite possible to obtain a textbook on electronic circuits written for engineers and one written for technicians. A text on electromagnetic theory can then be considered the scientific counterpart to these two.

However, recent technological advances do not fare as well as the more traditional areas of study. Robotics, electro-optics, lasers and fiber optics are areas that, while fairly well established, are not supported by a wide range of educational materials geared towards technicians, technologists, or undergraduates in engineering programs. Whereas the market certainly drives the product, it has been my experience that faculty who teach in these areas are ravenous for material to support excursions into these latest advances. This book, which is one of a very few of its kind, will address this need for technical information covered at a less theoretical (and mathematical) level and with more applied pedagogy.

   With these thoughts in mind, I have tried to compose a book that covers all the major areas of fiber optics, but leaves out high-level mathematics. At the same time, I have provided an adequate amount of theory and mathematics to explain the topics and allow for further exploration if a student so desires. The balance is, admittedly, a difficult one, and I would be happy to hear from any readers about how it may be improved.

   As a service to instructors using this text, I may point out that some elements (sections of Chapter 6 and 7 and all of Chapter 8, for example) can be omitted for courses that are composed of electronics students since they cover topics that are normally discussed in-depth in other courses. This book is designed to comfortably fill a normal fifteen- or sixteen-week semester, although I have managed to cover the same topics in a 10-week quarter with advanced students.

   Finally, I would like to acknowledge the reviewers of the manuscript for this text:

Robert J. Borns, Purdue University

Mazharul Haque, Texas State Technical College

Carl A. Jensen, Jr., DeVry Institute, Inc.

Jeffrey L. Rankinen, Pennsylvania College of Technology.

R.A.S.

✳ CONTENTS

AN INTRODUCTION TO FIBER OPTICS

1

AN INTRODUCTION TO OPTICAL FIBER

Included in this chapter:

1.1 THE NATURE OF OPTICAL FIBER

Optical fiber is a solid strand of glass (or, in some cases, plastic) that conducts light in much the same way that copper wire conducts electricity or pipes conduct water. Light travels through the fiber by reflecting from the fiber's inner surfaces. Because of the construction of the fiber, light that passes into the glass does not pass back out, but reflects and stays in the fiber. The fiber is very thin and flexible, and, therefore, it can be routed around corners and through small openings. Light passing through the fiber can be used for illumination, for sensing changes in temperature or other information, or can be used for sensing, welding, or cutting (for example, the high-intensity light in laser beams). However, its most common and most useful function is for communications.

   Communication is the process of conveying information from one place to another. In electronic communications, this is accomplished by imposing the information (signal) onto some form of energy (carrier) through modulation. The modulated carrier passes through a material (medium) and arrives at its destination where the information is removed through demodulation. A simple example is a telephone system in which a person's voice is the signal used to modulate the carrier (electricity). The carrier passes through the medium of copper wire and then is demodulated and transmitted through a speaker of another telephone.

In today's information age, the need for fast, accurate, and high-volume communication is paramount. Electronics and computers have permeated society so that they touch every aspect of our daily lives in some way. Bar code scanners read the prices of our grocery items, optical disks play our music, computers dispense money from automatic tellers, and robots controlled by computers assemble our manufactured items. We can use the personal computer for banking, real estate, accessing news and entertainment, and a hundred and one other functions that are an integral part of our life. This increase in electronic methods for manipulating, interpreting, and communicating information has revealed the limits of traditional communication technologies such as copper wire and radio waves. Optical fiber has been developed to overcome these disadvantages.

Optical fiber is a riverbed along which the huge stream of data generated by modern technology flows. The bed is wide, deep, and capable of handling all the data we can pour into it without overflowing its banks. Originally, fiber was used only for long-distance telephone lines where the need for its greater capacity outweighed its expense and difficulties. As the technology developed and its cost dropped, the fiber became viable for ship and airplane systems, medium- and short-haul telephone lines, local area networks, and even cable TV. This widespread use of optical fiber has occurred because of its advantages over electrical signals in copper wire.

1.2 ADVANTAGES AND DISADVANTAGES OF OPTICAL FIBER

Optical fiber has several advantages over traditional transmission methods (see Table 1.1). Light has a much higher frequency than radio waves or modulated signals in copper wire. Because of this high frequency, light has a very high information-carrying capacity. Optical fiber can transmit this high-frequency carrier with a small amount of distortion or loss of power. Therefore, optical fiber is capable of carrying hundreds and even thousands of times as much information as copper wire. Many more normal telephone conversations can be routed through a single fiber (systems capable of transmitting over 300,000 telephone calls through a single fiber have been demonstrated), as can data that require more capacity such as computer communications, video signals, and even a combination of signal types.

Table 1.1 Advantages of Optical Fiber

Property	Advantage	Application
Nonelectrical carrier	Electromagnetic immunity No electrical shock hazard	All
No electromagnetic emissions	Secure communications	All
High information-carrying capacity	Large amount of data High-speed transmission	Long distance Computer communications
Lightweight and small size	Small space requirement	Airplanes and ships Building systems Crowded city systems

Optical fiber cable is much smaller and lighter than copper cable, and it does not rust or corrode. This makes it invaluable in factories, airplanes, ships, buildings, and crowded urban areas where space and weight are a premium. The advantage of size is also apparent in long-distance communication where land is needed for laying the cable.

Because the fiber does not use electricity, it is not susceptible to noise from electronic equipment, sun spots, and other sources of electromagnetic energy that cause problems with radio and TV signals as well as signals transmitted through copper wire. Optical fiber can be routed near heavy equipment, computers, and other electrical devices without fear of crosstalk or noise.

Optical fiber does not emit any noise that might interfere with other signals or might be picked up by eavesdroppers. Signals sent through copper wire radiate electromagnetic energy that can be picked up by powerful antennas or other devices. The only way to eavesdrop on a signal as it passes through the optical fiber is to actually splice into the fiber, and even this is not nearly as simple as splicing into copper wire.

Although fiber optic's advantages outweigh its disadvantages in most cases, the disadvantages should be considered. Perhaps the most significant of these is that optical fiber requires a new set of skills for installation and maintenance by technicians. Technicians who are experienced in splicing, soldering, and installing copper wire will find optical fiber requires additional knowledge and training. Splicing and connecting optical fiber is a delicate skill that must be carefully learned to produce quality results (splices and connectors are discussed fully in Chapter 9).

Using optical fiber also means acquiring some specialized equipment. Volt-ohm-ammeters, oscilloscopes, and other devices used in testing and measuring electronic signals are replaced by optical power meters and optical time domain reflectometers. The methods and techniques for performing the tests and measurements are also changed (see Chapter 10).

These disadvantages mean that companies that are actively working in optical fiber must invest in equipment and training (or hire another company who already has). However, these challenges also offer opportunities for students studying to be technicians. Technicians trained in the methods of using, testing, and installing optical fiber will find many employment possibilities.

1.3 HISTORY OF OPTICAL FIBER

The use of fiber optics in communications has a fairly short history beginning in the late 1970s, but the principles of light in communications have been around much longer. The idea of using light to communicate evolved from simple signal fires and lamps (such as Paul Revere's famous "One if by land, two if by sea") to using mirrors to reflect the sun and transmit messages. The first modern attempt could be attributed to Claude Chappe and his optical telegraph built in France in the 1790s and to Alexander Graham Bell and his photophone patented in the 1880s (shown in Figure 1.1). Both devices relied on direct, line-of-sight transmissions of light. Bell's invention also incorporated the principles of modulating and demodulating the light.

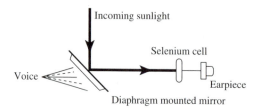

Figure 1.1 Alexander Graham Bell's Photophone

Bell, Chappe, and other early inventors and scientists were hampered by a lack of a reliable light source and the reliance on light passing through the air. The first problem was alleviated with the advent of the semiconductor age, the invention of the semiconductor laser by Theodore Maiman in 1962, the light emitting diode (LED), and the invention and improvement of optical fibers, which solved the problem of passing light through air.

Light communications through optical fiber was proposed in 1966 by Charles Kao and Charles Hockham of the Standard Telecommunication Laboratory in England. In 1970, Robert Maurer of Corning Glass Works produced the first fiber with loss under 20 dB/km. These efforts were preceded by development of an image-carrying bundle of fibers in the United States in 1951 and the introduction of fibers with cladding by Narinder Singh Kapany in 1953.

By 1978, telecommunications companies such as General Telephone Co. of Indiana, United Telephone Co. of Pennsylvania, and Cablecom General were field-testing fiber cable that were prototypes of the type of cable used today. Use of the fiber progressed rapidly through the 1980s with installation of cable within plants, under water, and along long-distance telephone lines. Improvements in technology brought more and more companies into the field. The rapid advances in optical fiber technology are still continuing.

Like so many of the new technologies introduced in the past few decades, optical fiber communication has become a well-developed and well-established technology in a short period of time. The use of optical fiber is definitely widespread, and its applications in communications as well as other fields are numerous.

1.4 MANUFACTURING OPTICAL FIBER

Optical fiber manufacturing is a well-established industry, and, like most modern manufacturing, it is a fairly complex process. The steps involved in manufacturing fiber can be simplified so that the process can be understood in general terms. The process has two steps, making the preform and drawing. The nature of fiber construction should be explained before these two steps are described.

Optical fiber used in communications is a solid glass strand (called the **core**) which is covered with another layer of glass (called the **cladding**). The fiber may have a distinct boundary between the core and cladding, or there may be a gradual change in materials between the two (more on this in Chapter 5). The dimensions of the fiber (diameter of core and cladding) vary depending on the application of the fiber, but these dimensions must be kept within tight specifications for a particular type of fiber. The type of glass used will also vary with application, but the purity (lack of any foreign materials as well as lack of imperfections such as cracks or bubbles) must also be kept very high. To produce quality fiber with these criteria in mind, manufacturers begin by making a preform.

A **preform** is a rod of glass containing the materials of the fiber and constructed the same way as the fiber but with much larger dimensions. The preform may be constructed by using the Modified Chemical Vapor Deposition (MCVD) process which uses a hollow tube and a stream of heated materials (known as the soot) passing through its center. The soot is used to form the inner part (or core) of the fiber, and it sticks to the walls of the tube. A torch follows along the outside of the tube and slightly behind the flow of soot and sinters the soot to the tube. Several passes may be run depending on the intended result. After the layers of soot are deposited, the torch moves along the tube at a slower rate and causes the tube to melt and collapse so that it forms a solid rod. The process is illustrated in Figure 1.2.

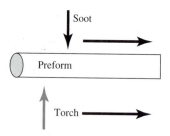

Figure 1.2 Making the Preform

The other process for making a preform involves depositing soot on the outside of a solid glass rod (known as the **bait rod**). The soot is deposited layer by layer, starting with the core material and moving on to the cladding material. Once the necessary soot is deposited, the bait rod is removed from the center. The hole left by the bait rod is eliminated in the drawing step.

After the preform is made, it is drawn out to form the fiber. As shown in Figure 1.3, the preform is placed into a mechanism that is shaped somewhat like a funnel. At the neck of the funnel, heat is ap-

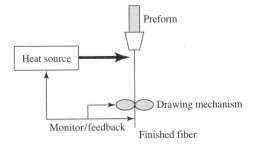

Figure 1.3   The Drawing Process

plied to the preform, and then the molten end is pulled through the neck, producing the fiber. The size of the fiber is monitored as it is pulled through the neck, and adjustments are made to the temperature of the preform and the speed of the drawing so that the size of the fiber is constant.

CHAPTER REVIEW

New Terms

Section 1.1
   Optical Fiber
   Communications
   Carrier
   Signal
   Medium

Section 1.3
   Photophone

Section 1.4
   Preform
   Modified Chemical Vapor Deposition
   Bait Rod

Other Chapters with Related Information

   Chapter 5
   Chapter 9
   Chapter 10

Review Questions

1. List the advantages of optical fiber and some applications where they might be useful.
2. Why does optical fiber communications have such a high information-carrying capacity?
3. Describe two methods for creating a preform.
4. Who invented the Photophone?
5. When was the semiconductor laser invented?
6. What is soot? How is it used to manufacture optical fiber?
7. Why did earlier attempts to communicate with light fail?
8. Describe an optical fiber.
9. What are some applications of optical fiber?
10. What are some companies that began field-testing optical fiber cables in 1978?

Thought Questions

1. What are some applications available now or in the future that might need the advantages of optical fiber?
2. What would make optical fiber a better choice than satellite communications? Microwave communications?
3. What role would optical fiber play in the information superhighway and the Internet?

REFERENCES

Cherin, Allen H. *An Introduction to Optical Fibers.* New York: McGraw-Hill, 1983.

Senoir, John M. *Optical Fiber Communications—Principles and Practice.* Upper Saddle River, NJ: Prentice Hall International, 1985.

Sterling, Donald J. *Technician's Guide to Fiber Optics.* Albany, NY: Delmar Publishers, Inc., 1987.

Zanger, Henry and Zanger, Cynthia. *Fiber Optics—Communications and Other Applications.* Upper Saddle River, NJ: Merrill/Prentice Hall, 1991.

2

✳ THE PHYSICS OF LIGHT

Included in this chapter:

2.1 INTRODUCTION

The theories and models of light are an integral part of the study of fiber optics. Very often, students of optical fiber technology come from an electronics background and are unfamiliar with the terms and concepts used by physicists and optical engineers who study the nature of light. One of the goals of this chapter is to help such students achieve a better understanding of light.

Traditionally, the study and applications of light are divided into three broad categories: **physical optics**, **geometrical optics**, and **quantum optics**. In each category, light is studied with a different model. Physical optics use the wave model (similar to radio and TV waves). Quantum optics uses a particle, or photon, model; geometrical optics considers light as a straight line ray. Each model is valid within limitations. The complete model of light involves all three. We will discuss only those elements necessary for understanding the topics in this book.

Physical optics (light as a wave) is the primary tool used to describe the phenomena of interference, diffraction, polarization, and birefringence. It is the basis for understanding applications of light such as holography, diffraction gratings, and optical coatings. In fiber optics, physical optics principles help to describe the propagation of light in modes (discussed in Chapter 5) and to explain coherent communications and polarization-maintaining fibers. Diffraction is also important in studying optical fiber.

Geometrical optics is used to define the rules that govern reflection and refraction. Devices such as mirrors, lenses, and splitters can be described using geometrical optics. Optical fiber systems also use these devices to focus and manipulate light before launching it into or after it escapes from the fiber. Refraction, the basic phenomena that describes light traveling through a fiber, is also an important element of fiber optics.

In studying quantum optics, light is considered as a particle, or photon. The principles of semiconductor devices used in fiber optics, such as the light-emitting diode, the semiconductor laser, and various optical detectors, can be explained using this perspective.

This chapter reviews all three fields of study, emphasizing those concepts that are of most interest to fiber optics. Complete discussions on the concepts introduced here can be found in the references at the end of the chapter.

2.2 PHYSICAL OPTICS

Light is a form of electromagnetic radiation. Like radio waves, microwaves, and other familiar waves used in communications, light is composed of two varying fields—an electric field and a magnetic field (see Figure 2.1). These two fields induce each other and allow light to propagate. The wave model of light is normally simplified to consider a single wave (rather than both the electric and magnetic fields). This simplification leads to the wave equation

$$Y = A \sin(kx - \omega t + \delta) \tag{2.1}$$

This equation describes a sine wave similar to the wave studied in alternating current (AC) electronics, and it can be used to determine the common properties of waves. The amplitude (A), the wave number (k), the initial phase angle (δ), and the angular frequency (ω) describe the fundamental properties of the wave although the wave number and angular frequency are generally converted to wavelength and frequency for most wave calculations.

The wave number (k) is related to the wavelength (λ) by the equation

$$k = 2\pi/\lambda \tag{2.2}$$

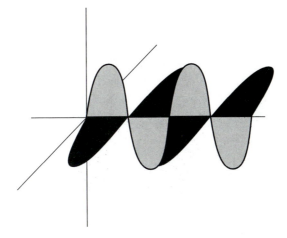

Figure 2.1 Electromagnetic Radiation

and the frequency (f) is related to the angular frequency (ω) by the equation

$$\omega = 2\pi f \qquad \qquad \textbf{(2.3)}$$

Wave number and angular frequency are useful for writing the wave equation in its simplest form, but the wavelength and the frequency represent actual, physical properties of the wave and are used more often in wave calculations. Sometimes it is necessary to derive the wavelength and frequency from a wave equation using the method shown in Example 2.1.

 EXAMPLE 2.1

Given the wave represented by the following equation, determine the wavelength and the frequency.

$$Y = 10 \sin(6\pi x - 4pt + 30)$$

To find the wavelength

Using Equation 2.2,

$$k = 2\pi/\lambda$$
$$\lambda = 2\pi/k$$

the wave number, k, is the coefficient of x in the wave equation, so the wavelength is calculated by

$$\lambda = 2\pi/6\pi = 1/3 \text{ meters}$$

To find the frequency

Using Equation 2.3,

$$\omega = 2\pi f$$
$$f = \omega/2\pi$$

The angular frequency (ω) is the coefficient of t in the wave equation, so that frequency is calculated by

$$f = 4\pi/2\pi = 2 \text{ hertz}$$

2.2.1 Amplitude, Wavelength, Frequency, and Velocity

The basic wave properties are summarized in Table 2.1, which also indicates the symbols commonly used to represent them and the base unit used in their measurement. Each wave property is described in detail in the following sections.

The **wavelength** (λ) is the distance between two like points along the wave, and is usually measured between two peaks. The **amplitude** (A or E) is the height of the wave and is similar to the peak voltage of an AC signal. Both of these properties are illustrated in Figure 2.2.

Table 2.1 Basic Wave Properties

Wave Property	Symbol	Base Unit
wavelength	λ	meters
frequency	v or f	hertz
amplitude	A or E	
phase	δ	radians or degrees
velocity	v	meters/second

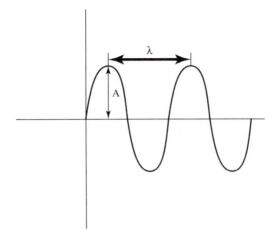

Figure 2.2 Amplitude (A) and Wavelength (λ)

The **frequency** and velocity of the wave arise from the motion of the wave. Since the wave is moving, it has a speed or velocity associated with it. The number of wavelengths that pass a fixed point in a second is known as the *frequency*. The concepts of velocity, frequency, and wavelength can be envisioned by considering a train passing by an intersection. The cars of the train have a certain length (corresponding to wavelength), and the train is traveling at a certain velocity. If you count the number of train cars that pass you in one second, you are measuring the train's frequency.

The train analogy also illustrates the relationship between velocity, wavelength, and frequency. The frequency (number of train cars that pass) is affected by the velocity. If the train moves faster, more cars pass in one second; if the train moves slower, fewer cars pass. In other words, frequency and velocity are directly proportional.

Frequency is also affected by wavelength. If each train car is shorter, more cars are able to pass in one second; if each car were longer, fewer cars could pass. The wavelength and the frequency are indirectly proportional. The relationships between wavelength, frequency, and velocity are expressed mathematically as

$$v = f\lambda \qquad\qquad (2.4)$$

For most computations, the velocity of light is assumed to be a constant corresponding to the velocity of light in a vacuum, 3×10^8 m/s, and referred to with the letter c.

EXAMPLE 2.2

Given electromagnetic waves in a vacuum, determine the wavelength for the following frequencies: 150 GHz, 225 MHz, 14 THz.

To find the wavelength

In a vacuum, the velocity of the waves is 3×10^8 m/s. Using Equation 2.4,

$$v = c = 3 \times 10^8 \text{ m/s} = f\lambda$$

Rearranging Equation 2.4 to solve for wavelength (λ) gives

$$\lambda = (3 \times 10^8 \text{ m/s}) / f$$

The wavelengths for the given frequencies are then

$f = 150$ GHz
$\lambda = (3 \times 10^8 \text{ m/s}) / 150 \text{ GHz} = 2.00$ mm
$f = 225$ MHz
$\lambda = (3 \times 10^8 \text{ m/s}) / 225 \text{ MHz} = 1.33$ m
$f = 14$ THz
$\lambda = (3 \times 10^8 \text{ m/s}) / 14 \text{ THz} = 21.4$ mm

As Example 2.2 shows, waves with high frequencies have short wavelengths, whereas waves with low frequencies have long wavelengths. The difference between various types of electromagnetic waves lies in the difference in their wavelength (or frequency). Light waves are high frequency (short wavelength), and radio and TV waves are low frequency (long wavelength). A list of all types of electromagnetic waves in order by frequency or wavelength is known as the **electromagnetic spectrum** (illustrated in Figure 2.3)

Notice that the electromagnetic spectrum is arranged by both wavelength and frequency. For most areas of optics, the wavelength is used in describing the wave.

The light waves that are visible to the human eye are roughly in the 400 nm to 700 nm range, but there are two other ranges that are not visible but are usually referred to as light. From roughly 3 nm to 400 nm lies ultraviolet light, and from roughly 700 nm to 10,000 nm is infrared light. The wavelength (or frequency) of visible light corresponds to the color of light. Green light is around 500 nm, red, 600 nm, and violet, 400 nm.

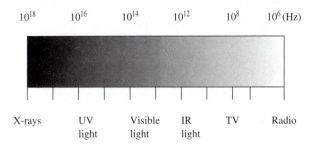

Figure 2.3 Electromagnetic Spectrum

Light occurs naturally as a combination of several different waves, each with a separate wavelength. Even light that we think of as one color (the red light on a police car, for example) is actually several waves whose wavelengths are close in value. When light is traveling, each wavelength travels at a slightly different velocity. As a result, the velocity of light in a material can be described in one of two ways. **Group velocity** refers to the velocity of all the waves traveling together. **Phase velocity** is the velocity of a single point of constant phase in the wave.

2.2.2 Phase and Coherence

Light sources generally emit multiple waves, and the relationship among these waves is described in terms of phase. **Phase** indicates the position of one wave relative to another. The difference in position between two waves is measured as the phase difference, expressed in degrees or radians. Two waves that are aligned, as shown in Figure 2.4(a), are commonly known as **in-phase** or are said to have a phase difference of zero degrees (0 radians). The two waves shown in Figure 2.4(b) are **out of phase** by 180 degrees (π radians).

The phase relationships of waves from a source of light are often described by the term **coherence**. Light that has coherence (or coherent light) has a fixed phase among the waves. In other words, all of the waves of a coherent light source are aligned the same way. Incoherent light waves have a phase that is constantly changing.

Coherence is also generally intended to mean that all of the waves are of the same wavelength (also known as **monochromatic**). To clarify the terms, light is sometimes described as **spatially coherent** (waves have a fixed phase) and **temporally coherent** (waves are monochromatic, i.e., have the same wavelength).

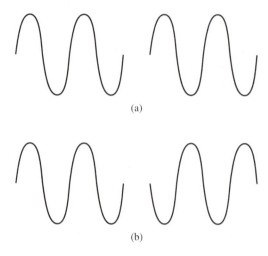

(a)

(b)

Figure 2.4 (a) In-Phase Waves (b) Out-of-Phase Waves

2.2.3 Wave Phenomena

Light waves exhibit some distinct phenomena which are an important part of many applications. Foremost among these phenomena are interference, diffraction, and polarization. Understanding each of these is important in studying light.

Interference is an effect observed under special conditions. When coherent waves are combined, they exhibit one of two effects. If the light waves are 180 degrees out of phase, the sum of the two waves will be a third wave with an amplitude equal to the difference between the amplitudes of the individual waves. If the two waves that are combined have the same amplitude, the result will be no wave at all. This effect is known as **destructive interference**.

Constructive interference is when light waves are in phase with each other, resulting in a third wave whose amplitude is equal to the sum of the amplitudes of the interfering waves. Constructive and destructive interference are illustrated in Figure 2.5.

Interference is closely related to another wave effect known as **diffraction**. When light passes through a small opening, it spreads out to fill areas that would be expected to be in a shadow. As shown in Figure 2.6, diffraction causes a change in the shape of the incoming light. The exact effect of diffraction depends largely on the size of the opening, the distance between the light source and the opening, and the wavelength.

Diffraction is classified as either Fraunhofer or Fresnel. **Fraunhofer** (or far field) **diffraction** occurs when the light source, the opening, and the observation point are large distances away. **Fresnel** (or

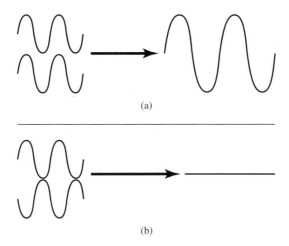

(a)

(b)

Figure 2.5 (a) Constructive Interference (b) Destructive Interference

near field) **diffraction** occurs when a light source and observation point are near the opening.

The concept of diffraction has been used to construct useful optical devices such as **diffraction grating** which can be used to separate light into its component wavelengths. Diffraction is also the principle used to describe the adverse effects of emitting light through a small opening. Light sources used in fiber optics are generally of the semiconductor type (see Chapter 7 for a complete discussion). These sources emit light from a small opening at the junction between two semiconductor materials, and, as a result, the emitted light is diffracted or spread out.

Because light spreads out from a semiconductor source, it is difficult to inject it into the small core of a fiber. The effects of diffraction must be limited to make launching light into the fiber feasible. Diffraction effects are partially tempered by the construction of the light source, and they can be further limited by placing optical components between the source and the fiber.

Polarization is the last major concept in wave optics. The polarization of a wave is really just an expression of its orientation in space. Polarization should not be confused with phase which describes orientation of a wave relative to another wave. For a typical source of light, the waves emitted are randomly oriented. Some waves oscillate vertically, while others oscillate horizontally, and still others oscillate at some angle in between. This haphazard distribution of orientations is normally known as **random polarization** (or unpolarized light).

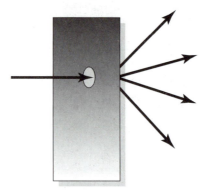

Figure 2.6 Diffraction

Some types of light sources, and some optical effects, produce light with a set polarization where all the waves are oriented in the same direction. A good example is reflection, in which reflected light waves are all polarized in a direction parallel to the surface of reflection. In fiber optics, reflected light causes loss in signal strength and can cause noise in the system. The polarizing effect of reflection is used to design devices with minimal reflection effects.

2.3 QUANTUM OPTICS

The principles of quantum optics are based on the theories of quantum mechanics and the structure of the atom. The atom, commonly described as the building block of matter, consists of a central structure called the nucleus which is surrounded by orbiting electrons. The system is somewhat similar to our solar system where the sun is orbited by planets. The nucleus of an atom is composed of two types of particles: the proton, which has a negative charge, and the neutron, which has no charge. The electrons orbiting these particles have a negative charge. The energy levels of these electrons is the key to the production of light.

Atoms may contain varying amounts of energy depending on their structure. The amount of energy that an atom may contain is quantitized, or limited to specific and discrete amounts. To illustrate this point, consider the makeup of the simplest atom—the hydrogen atom. As shown in Figure 2.7, the hydrogen atom has a nucleus containing a single proton (no neutrons). Orbiting this nucleus is a single electron. The electron travels around the nucleus, similar to a planet orbiting the sun, but the electron is able to follow one of several possible paths (il-

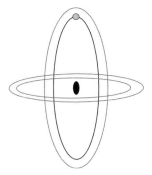

Figure 2.7 The Hydrogen Atom

lustrated by the dotted circles in Figure 2.7). The path (or orbital) the electron chooses depends on the energy state of the atom.

The important aspect of the electron orbit is that, although the electron has several possible orbitals, it is confined to these orbitals and these alone. The electron may use orbitals $n = 2, 3, 4$, etc., but it cannot exist in the areas between. The position of the electron is therefore quantitized.

Since the orbital used by an electron depends on the energy state of the atom, it follows that these energy states must also be quantitized. The atom can, therefore, contain certain specific levels of energy which correspond to certain specific electron orbitals. If energy is supplied to the atom, it will be absorbed in these specific amounts; if energy is released from the atom, it will be released in the same specific amounts.

Exchange of energy by an atom is important to optics since this is how light is produced. Consider an atom which contains its minimum amount of energy (called a **ground state** atom). Its electron(s) is orbiting close to the nucleus, but if energy were absorbed by this atom, its **valence** (outermost) **electron** would be moved to an orbital farther from the atom. After absorbing the energy, the atom is said to be in an **excited state**.

Atoms do not remain in excited states for very long. After a short time, the atom returns to its ground state by releasing the energy it has absorbed. This released energy is in the form of a particle or packet called a **photon**, which, depending on the atom and energies involved, could be some form of light. The wavelength of light produced in this manner depends on the amount of energy that the atom absorbed and then released. This relationship is governed by the equation

$$E = h\lambda \tag{2.5}$$

where E is the energy, h is a constant (known as **Planck's Constant**), and λ is the wavelength.

The relationship between wavelength and energy becomes more significant when it is realized that the energy absorbed and emitted by a particular atom is limited to certain discrete values. Also, the energy values are different for different types of atoms (for example, helium has one set of energies whereas sodium has another). Because of this, we conclude that each type of atom will absorb and emit its own unique group of wavelengths.

Table 2.2 shows some of the wavelengths emitted by common atoms. These emissions can be observed in everyday life by looking at the emissions of known atoms. For example, neon, used in neon signs, emits several bright colors such as red, green, and orange.

2.4 GEOMETRICAL OPTICS

The field of geometrical optics, founded on the principle of **rectilinear propagation** (light travels in a straight line), uses lines (or rays) to illustrate the path that light follows. Two basic concepts are defined in geometrical optics: reflection and refraction. The rules governing the effects of these concepts on the direction that light travels are the fundamental rules of geometrical optics.

2.4.1 Reflection and the Law of Reflection

Reflection occurs when light traveling in one material reaches the surface of another material and bounces off. The light may either completely reflect or part of it may reflect with part passing into the other

Table 2.2 Some Emission Wavelengths of Common Atoms

Atom	Emission Wavelength (nm)	Brightness
Mercury	404.66 (violet)	Faint
	546.07 (green)	Bright
	576.96 (yellow)	Bright
Helium	447.15 (blue)	Bright
	501.57 (blue-green)	Bright
	667.2 (red)	Medium
Neon	632.8 (red)	Bright
	543 (green)	Faint

material, depending on the materials involved and the wavelength of the light. Reflection is categorized into three types: fresnel, specular, and diffuse.

In **Fresnel reflection**, light is incident with the surface of a transparent or semitransparent surface (such as glass). Most of the light passes into the new material, but a small percentage (usually about 4%) reflects.

In **specular reflection**, light reflects from a highly reflective surface (such as a mirror), and a large percentage (generally close to 100%) reflects.

Diffuse reflections come from a rough, opaque surface (such as a piece of wood). The amount of light that reflects is variable, but the reflected light scatters in several directions, unlike specular and Fresnel reflection.

All reflections are governed by the law of reflection. This law states that the angle at which the light approaches a reflective surface is equal to the angle at which it reflects. As illustrated in Figure 2.8, the angles involved are measured relative to a line (called the normal to the surface) drawn perpendicular to the reflective surface. Using the quantities defined in Figure 2.8, the law of reflection is expressed mathematically as

$$\theta_i = \theta_r \tag{2.6}$$

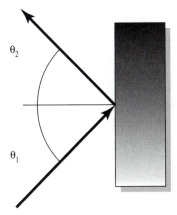

Figure 2.8 Reflection

2.4.2 Refraction and Snell's Law

When light passes from one material to another, it changes the direction in which it is traveling. This bending of light is known as **refraction**. The amount and direction of the bend can be determined by knowing the index of refraction of the materials involved.

Index of refraction is a measure of the speed that light will travel through a certain material. Recall from Section 2.2.1, that the speed of light in a vacuum is 3×10^8 m/s. When light travels through any material (such as glass or plastic), its speed is less than its speed in a vacuum. The index of refraction (n) of a material is calculated from the ratio of the speed of light in a vacuum to the speed of light in the material.

$$n = (3 \times 10^8 \text{ m/s})/(\text{speed in the material}) \quad\quad (2.7)$$

Table 2.3 lists the index of refraction for some common materials along with the speed light travels in them.

Table 2.3 Index of Refraction for Common Materials

Material	Index	Speed (m/s)
air	1.000292	2.991×10^8
water	1.333333	2.250×10^8
sodium chloride	1.54	1.948×10^8
diamond	2.42	1.181×10^8

 EXAMPLE 2.3

Given the speed of light in the following types of glass, calculate the index of refraction for each one.

Barium Flint: 1.89×10^8 m/s
Spectacle Crown: 1.97×10^8 m/s
Fused Quartz: 2.06×10^8 m/s

To find the index of refraction
Using Equation 2.7,

$$n = \frac{(3 \times 10^8 \text{ m/s})}{(\text{speed in the material})}$$

Barium Flint

$$n = \frac{3 \times 10^8 \text{ m/s}}{1.89 \times 10^8 \text{ m/s}} = 1.59$$

Spectacle Crown

$$n = \frac{3 \times 10^8 \text{ m/s}}{1.97 \times 10^8 \text{ m/s}} = 1.52$$

Fused Quartz

$$n = \frac{3 \times 10^8 \text{ m/s}}{2.06 \times 10^8 \text{ m/s}} = 1.4$$

The index of refraction is used to calculate the angles involved in refraction with the help of an equation known as **Snell's Law**. Consider the diagram in Figure 2.9. Snell's Law states that the index of refraction of the two materials and angles involved are related by the equation,

$$n_1 \sin \theta_1 = n_2 \sin \theta_2 \qquad \textbf{(2.8)}$$

where the subscripts 1 and 2 refer to the angles and the indexes of the two materials.

Snell's Law can be used to calculate the angle at which light will refract, the indexes of the material, or the incident angle required to produce a certain refracted angle. Some of the problems that can be solved using Snell's Law are given in Example 2.4.

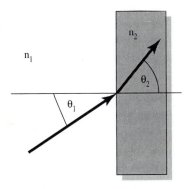

Figure 2.9 Snell's Law

EXAMPLE 2.4

a. Light travels from air ($n = 1.00$) to water ($n = 1.33$) at an incident angle of 40°. Calculate the refracted angle.

To find the angle

Using Equation 2.8,

$$n_1 \sin \theta_1 = n_2 \sin \theta_2$$

divide both sides by n_2

$$n_1/n_2 \sin \theta_1 = \sin \theta_2$$

Take the arcsine of both sides

$$\theta_2 = \text{arcsine}(n_1/n_2 \sin \theta_1)$$
$$= \text{arcsine}(1.00/1.33 \sin (40)) = 28.9°$$

b. What index should a material have if it were used in place of the air in part (a) to produce a refracted angle of 50°?

To find the index

Using Equation 2.8,

$$n_1 \sin \theta_1 = n_2 \sin \theta_2$$

divide by $\sin \theta_1$

$$n_1 = n_2 \sin \theta_2/\sin \theta_1$$
$$= 1.33 \sin (50)/\sin (40)$$
$$= 1.59$$

Example 2.4 shows that when light travels from a material with a larger index of refraction to a material with a smaller index of refraction, it refracts away from the normal. Light traveling from a material with a smaller index to a material with a larger index refracts toward the normal. In the first case (larger index to smaller index), increasing the angle of incidence increases the angle of refraction to the point where the light actually reflects instead of refracts. The angle of incidence (θ_c) that produces this reflection can be calculated by

$$\theta_c = \text{arcsine}(n_2/n_1) \tag{2.9}$$

and is called the **critical angle**.

The critical angle is important to optical fiber. The fiber is constructed so that the index of refraction of the core is larger than the index of cladding. Light is injected into the fiber so that it is incident to the core/cladding intersection at the critical angle. The light passes through the fiber by repeating this reflection effect.

 EXAMPLE 2.5

Calculate the critical angle between two materials with indices of $n_1 = 1.45$ and $n_2 = 1.40$.

Critical angle is given by Equation 2.9

$$\begin{aligned} \theta_c &= \text{arcsine}(n_2/n_1) \\ &= \text{arcsine}(1.40/1.45) \\ &= 75° \end{aligned}$$

CHAPTER REVIEW

New Terms

Section 2.1
    Physical Optics
    Geometrical Optics
    Quantum Optics

Section 2.2
    Amplitude
    Wavelength
    Frequency
    Phase
    Coherence
    Interference
    Polarization

Section 2.3
   Orbital
   Excited State
   Ground State
   Quantitized

Section 2.4
   Reflection
   Refraction
   Index of Refraction
   Snell's Law
   Critical Angle

Other Chapters with Related Information
   Chapter 3
   Chapter 5
   Chapter 7

Review Questions
   1. What is the difference between physical optics, geometrical optics, and quantum optics?
   2. Describe diffraction.
   3. Describe interference.
   4. Describe polarization.
   5. How are wavelength and frequency related?
   6. What does the index of refraction measure?
   7. What is a photon?
   8. Name three types of electromagnetic energy.
   9. What two conditions must be met for light to reflect rather than refract?
   10. Name three materials that might be diffuse reflectors.

Thought Questions
   1. How would diffraction affect light exiting from the end of an optical fiber? What could be done to reduce these effects?
   2. How could the concept of refraction be used in the design of a lens?

3. What type of reflection would be likely to occur in an optical fiber?
4. How would the choice of materials, and therefore the choice of index of refraction, affect the parameters of an optical fiber?
5. Why do different sources of light emit different wavelengths?

Problems

1. Determine the wavelength of electromagnetic waves with the following frequencies (assume the waves are in a vacuum): 100 GHz, 150 kHz, 10.4 MHz.
2. If an electromagnetic wave was described by the equation, $Y = 10 \sin (2\pi x - 10\pi t + 30)$, determine its wavelength and frequency.
3. Light travels from air into a diamond at an incident angle of 10°. Determine the refracted angle.
4. What would be the critical angle for light traveling from the diamond back into the air?
5. What is the energy of a 400-nm photon if $h = 6.6 \times 10^{-34}\,j\text{-}s$?

REFERENCES

Blaker, J. Warren and Rosenblum, William M. *Optics—An Introduction for Students of Engineering*. New York: Macmillan, 1993.

Hecht, Eugene, and Zajac, Alfred. *Optics*. Reading, MA: Addison-Wesley Publishing, 1979.

Jenkins, Francis, and White, Harvey. *Fundamentals of Optics*. New York: McGraw-Hill, 1976.

Kingslake, Rudolph. *Lens Design Fundamentals*. New York: Academic Press, 1978.

3

✳ LIGHT PROPAGATION

Included in this chapter:

3.1 INTRODUCTION

Several different methods for studying the propagation of light are available. Each method was developed to obtain specific information about some aspect of the propagation, and the methods tend to mirror the theories of light discussed in Chapter 2. In this chapter, we select information from different models for light propagation to explain more clearly how light travels through the fiber.

The propagation of light can be explained from several perspectives, ranging from complex mathematical models based on the principles of physical optics to simplistic, intuitive descriptions. Which perspective is best depends on what career path a student has chosen. Scientists and engineers may require a complex model to help them design or research fibers or fiber systems, whereas a technician may need an explanation that reveals the practical aspects of the fiber.

This chapter explains the fundamental concepts behind light propagation, but avoids, whenever possible, the complex mathematical models upon which they are based. This approach gives students a good working knowledge of light propagation and an understanding of the concepts that they can apply to their work with optical fiber. Students who are interested in learning more about the mathematical models should refer to the references at the end of the chapter.

In addition to discussing light propagation, this chapter provides information about how light is measured and categorized and the properties of the types of light normally used in optical fiber.

3.2 RAY PROPAGATION—
GEOMETRICAL OPTICS MODEL

One method for studying light propagation that does not involve complex mathematics is known as **ray tracing**. Based on the principles of refraction and Snell's Law, ray tracing is a method of plotting the path followed by a ray of light through an optical system. For optical fiber, the process of tracing rays is simplified because, once the light enters the fiber, the rays do not encounter any new surfaces but repeatedly hit the same surface (assuming, of course, the fiber has no flaws).

To understand ray tracing in a fiber, recall the principles of Snell's Law discussed in Chapter 2. Light traveling from one material to another will refract, or bend. The direction and the amount of refraction depends on the initial angle and the indexes of refraction for the two materials as stated in Equation 2.8 from Chapter 2.

$$n_1 \sin \theta_1 = n_2 \sin \theta_2 \qquad \textbf{(2.8)}$$

We also know that if light is traveling at the proper angle from a material with a higher index to one of a lower index, the light will reflect rather than refract. This process is known as *total internal reflection*, and the angle required is called the *critical angle*, which can be calculated from Equation 2.9 from Chapter 2.

$$\theta_c = \text{arcsine}(n_{\text{core}}/n_{\text{cladding}}) \qquad \textbf{(2.9)}$$

where the subscripts *core* and *cladding* have been used to describe the index of refraction of the core and the index of refraction of the cladding.

To understand how these two principles are related to optical fibers, consider Figure 3.1. Light enters the end of an optical fiber at some angle, θ_1.

Because the light is traveling from one material to another (air to the glass of the fiber), it will refract. From Equation 2.8, we know the new angle formed by the fiber, θ_1', will be equal to

$$\theta_1' = \text{arcsine}(n_{\text{air}}/n_{\text{core}}) \sin \theta_1 \qquad \textbf{(3.1)}$$

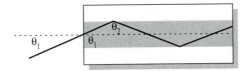

Figure 3.1 Snell's Law in a Fiber

Once the light has entered the fiber, it travels through the core to the surface of the cladding. Light is incident with the cladding at an angle of θ_2. From the principles of total internal reflection, we know that if θ_2 is greater than or equal to the critical angle, the light will reflect and remain in the core.

Since the goal of optical fiber is to contain light in the core, the light coming into the fiber should strike the cladding at the critical angle. Using Equations 2.8, 2.9, and 3.1, we can determine the value of θ_1 that will cause the light to reflect at the core/cladding interface. For the reflection to occur, θ_2 must be equal to or greater than the critical angle. From Equation 2.9,

$$\sin \theta_2 > (n_{core}/n_{cladding}) \qquad \text{(3.2a)}$$

To simplify the mathematics involved, we can remove the greater than sign in Equation 3.2(a) so that it becomes

$$\sin \theta_2 = (n_{core}/n_{cladding}) \qquad \text{(3.2b)}$$

The greater than sign can be reinserted after the final equation is derived. From Snell's Law (Equation 2.8), the sine of θ_1' is defined as

$$\sin \theta_1' = (1/n_{core}) \sin (\theta_1)$$

where the index of refraction of air is assumed to be 1.000. The geometry in Figure 3.1 indicated that the sine of θ_1' is equal to the cosine of θ_2, so we can write

$$\sin \theta_1' = \cos \theta_2 = (1/n_{core})(\sin \theta_1) \qquad \text{(3.3)}$$

Combining Equations 3.2(b) and 3.3 and using some basic algebra, we find that

$$\sin \theta_1 = \sqrt{(n^2_{core} - n^2_{cladding})} \qquad \text{(3.4a)}$$

We can now reinsert the greater than sign removed in the previous steps, which gives

$$\sin \theta_1 < \sqrt{(n^2_{\text{core}} - n^2_{\text{cladding}})} \qquad \textbf{(3.4b)}$$

The expression in Equation 3.4(b) states that as long as θ_1 is less than or equal to the square root term, light entering the fiber will reflect from the core/cladding interface and stay in the fiber. Input angles larger than those calculated by Equation 3.4(b) will cause light to refract and pass into the cladding.

The right side of Equation 3.4(b) is known as the **numerical aperture** of the fiber. The angle, θ_1, in Equation 3.4 is known as the **acceptance angle** of the fiber. Both values are used to describe the limits placed on light injected into the fiber, and they are often included in the specifications that fiber manufacturers provide to their customers. Example 3.1 illustrates how to calculate the numerical aperture and acceptance angle of a fiber.

 EXAMPLE 3.1

Calculate the numerical aperture and acceptance angle of a fiber with a core index of 1.54 and a cladding index of 1.50.

To calculate the numerical aperture (NA)

$$\begin{aligned} \text{NA} = \sin \theta_1 &= \sqrt{(n^2_{\text{core}} - n^2_{\text{cladding}})} \\ &= \sqrt{(1.54^2 - 1.50^2)} \\ &= 0.349 \end{aligned}$$

To calculate the acceptable angle

$$\begin{aligned} \theta_1 &= \sin^{-1}(\text{NA}) \\ &= \sin^{-1}(0.349) \\ &= 20.4° \end{aligned}$$

3.3 FIBER MODES—PHYSICAL OPTICS MODEL

The equations discussed in the preceding section seem to suggest that there are an unlimited number of possible input angles that will lead to light propagating through a fiber. However, due to the wave nature of light, the number of possible input angles that allow light to propagate through optical fiber is fixed by the dimensions of the fiber.

Wave theory dictates that waves confined by a cavity (such as an optical fiber) produce specific patterns. These patterns are a function of the wavelength and the dimensions of the cavity. For example, the wave shown in Figure 3.2 is described by the equation

$$Y_1 = A \sin (kx - \omega t) \tag{3.5}$$

As the wave propagates through the cavity, it will reach the wall of the cavity and reflect. The reflected wave will be described by

$$Y_2 = A \sin (kx + \omega t) \tag{3.6}$$

where the minus sign has been changed to a plus sign since the wave reverses directions after reflection.

When the two waves (the original and the reflected wave) meet, they will combine to form a resultant wave (Y_r) with the equation

$$Y_r = Y_1 + Y_2 = A[\sin(kx - \omega t) + \sin(kx + \omega t)] \tag{3.7}$$

Equation 3.7 can be simplified using some basic trigonometry to produce the formula,

$$Y_r = 2A \cos(\omega t) \sin(kx) \tag{3.8}$$

which describes the combination of the original wave and the reflected wave. Since this combined wave exists inside a cavity, it is confined by a set of *boundary conditions*. Boundary conditions are details of the physical process that is being described mathematically, and they are generally used to limit the equation.

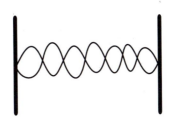

Figure 3.2 Standing Waves

In this example, the boundary conditions are that the wave must have a zero amplitude at the two ends of the cavity ($x = 0$ and $x = L$). If the conditions are not true, the wave would not exist as shown in Figure 3.2. Sometimes these specific boundary conditions are expressed by saying that the wave must be a *standing wave*.

Mathematically, the standing wave boundary conditions mean that Y_r must be equal to zero when $x = 0$ and when $x = L$,

$$Y_r(x = 0) = 2A \cos(\omega t) \sin(k \times 0) = 0 \qquad \textbf{(3.9a)}$$
$$Y_r(x = L) = 2A \cos(\omega t) \sin(k \times L) = 0 \qquad \textbf{(3.9b)}$$

Equation 3.9(a) is automatically correct since $\sin(k \times 0)$ is equal to zero, but Equation 3.9(b) requires that $\sin(k \times L)$ must be equal to zero or,

$$\sin(k \times L) = 0$$

or,

$$k = [\arcsin(0)]/L$$
$$= (j\pi)/(2L) \qquad \textbf{(3.10)}$$

where j is equal to 1, 3, 5, 7 . . ., etc. The limit placed on the wave number, k, by Equation 3.10 means that there is a limit on the wavelength (recall from Chapter 2 that wavelength and wave number are related). In practical terms, this means that for a certain cavity size, L, only specific wavelengths will form standing waves.

An optical fiber is also a cavity. Light propagating through the fiber is a wave; therefore, the standing wave limits in Equation 3.10 will also apply in a fiber. Since light in the fiber propagates at an angle to the core/cladding interface, the cavity size is governed by this angle. For a particular wavelength of light, only specific angles will produce standing waves. In other words, even though any angle that fits Equation 3.4 will cause light to reflect from the core/cladding interface and stay in the fiber, only specific angles will produce standing waves.

When light enters the fiber it is limited to the light that enters at an angle that will produce reflection (Equation 3.4) and will produce standing waves. The standing wave limitation is expresses mathematically as

$$\sin \theta_1 = j\lambda(2d) \qquad \textbf{(3.11)}$$

where $j = 0, 1, 2, 3$ . . . and d is the diameter of the core. Combining the constraints of Equation 3.11 and Equation 3.4 allows the calculation of which angles will actually lead light propagating through the fiber. Example 3.2 demonstrates how this calculation is made.

 EXAMPLE 3.2

Calculate which angles will lead to light propagation at a wavelength of 1300 nm in a fiber with a numerical aperture of 0.2, and with a core diameter of 100 μm.

To calculate the acceptance angle

Using Equation 3.4(b),

$$\theta = \sin^{-1}(NA) = \sin^{-1}(0.2) = 11.5°$$

Using Equation 3.11 and incremental values for j, calculate the angles that will form standing waves. Stop when you reach a value greater than the acceptance angle.

$$\sin \theta = j\lambda/2d$$

j	θ	j	θ	j	θ	j	θ
1	0.37	8	3.0	15	5.6	22	8.2
2	0.74	9	3.4	16	6.0	23	8.6
3	1.1	10	3.7	17	6.3	24	9.0
4	1.5	11	4.1	18	6.7	25	9.4
5	1.9	12	4.5	19	7.1	26	9.7
6	2.2	13	4.8	20	7.5	27	10
7	2.6	14	5.2	21	7.8	28	11

Example 3.2 illustrates that there can be several angles that meet the requirements for propagation. Each of these angles corresponds to a separate path through the fiber core that is called a *mode*. The number of modes in a fiber can be calculated based on a fiber parameter known as the *V number* (or *normalized frequency*). The *V* number is calculated by

$$V = (\pi d NA)/\lambda \tag{3.12}$$

where d is the core diameter and NA is the numerical aperture. For a large number of fiber modes, the V number is used to determine the number of modes (N) by

$$N = V^2/2 \qquad\qquad \textbf{(3.13)}$$

Modes in a fiber have several effects on the light that passes down the core. When light is initially injected into the fiber, it may not enter all of the possible modes (*underfilled* condition), or it may fill all of the possible modes and spill into the cladding (*overfilled* condition). After light propagates a fairly long distance in the fiber, light in the cladding will pass out into the air, and any empty modes will be filled so that eventually the fiber will reach **mode equilibrium**.

3.4 SKEW RAYS—
ALTERNATIVE PATHS

The light propagation discussed in Sections 3.1 and 3.2 is based on light that passes through the center of the fiber after each reflection from the core/cladding interface. This type of light is known as **meridional rays**. However, some light propagates through the fiber without passing through the center of the core. This light, known as **skew rays**, follows a helical (or spiral) path through the fiber as illustrated in Figure 3.3.

Skew rays do not obey the mathematical formulas developed for meridional rays. For example, the numerical aperture for skew rays is based on an acceptance angle and an angle (γ) which is the angle between the skew ray and the radius of the fiber. The skew ray numerical aperture (Na_s) is calculated by,

$$NA_s = (\sin \theta)(\cos \gamma) \qquad\qquad \textbf{(3.14)}$$

Figure 3.3 Skew Rays (End View of Fiber)

This equation shows that the numerical aperture of a skew ray is actually larger than that of a meridional ray. Light injected into a fiber outside of the normal acceptance angle may still propagate as a skew ray in the fiber.

Skew rays also offer an advantage when the light exits the fiber. The angle that the skew ray leaves a fiber does not depend on the conditions under which it entered the fiber as it does with meridional rays. Skew rays tend to smooth out the effects of these input conditions.

3.5 CHARACTERISTICS OF LIGHT

Light propagating through an optical fiber can be characterized in terms of several different parameters. Some of these parameters, such as wavelength and velocity, were described in Chapter 2, but there are many other methods for characterizing light. How the amount of light and the shape of a beam light can be described is important to the study of fiber optics.

The amount of light is most commonly described as the light power. Power is a measure of the amount of electromagnetic energy provided by a source of light in one second. The base unit of power is the watt, and, for most optical fiber applications, powers range from microwatts to milliwatts. Although a useful measurement, the power fails to describe some of the important aspects of light.

When light propagates through an optical fiber, a small percentage is lost through different mechanisms (these loss mechanisms are described more fully in Chapter 6). The loss of power is described using the **decibel** (dB). The decibel is a logarithmic measurement which allows measurement over a wide range of values. Since the decibel can be used to describe both an increase and decrease in power, loss is generally indicated with a negative sign. In other words, a gain in power may be 5 dB; a loss, −5 dB. In some measurements, it is assumed that a loss occurs so the negative sign is not used.

Decibels are calculated using the amount of power before the loss (P_o) and the amount of power after the loss (P). The number of decibels is calculated from

$$\# \text{ of dB} = 10 \log(P/P_o) \tag{3.15}$$

Sometimes, loss is compared to a predefined original power of 1 mW to provide a loss of value in dBm. In this case. P_o in Equation 3.15 is set to 1.

In addition to the loss of power, the area in which the power is distributed is important. This is especially true when the light exits the fiber and is detected. The detector's response depends on how much power is confined to the light-sensitive area of the detector.

The light power per area is known as the **irradiance** (*E*). The irradiance is calculated by dividing the power by the area of the detector, so irradiance units are usually watts per square centimeter or watts per square millimeter.

CHAPTER REVIEW

New Terms

Section 3.2
>   Ray Tracing
>   Numerical Aperture
>   Acceptance Angle

Section 3.3
>   Boundary Conditions
>   Standing Waves
>   Modes
>   *V* Number
>   Normalized Frequency

Section 3.4
>   Meridional Rays
>   Skew Rays

Section 3.5
>   Power
>   Decibels
>   dBm
>   Irradiance

Other Chapters with Related Information

>   Chapter 2
>   Chapter 6

Review Questions

1. What role does internal reflection play in light propagation through an optical fiber?

2. How is a skew ray different from a meridional ray?

3. What are some ways to describe the light traveling in the fiber?

4. Define numerical aperture and acceptance angle.

5. What are the modes in a fiber?

Problems

1. Determine the numerical aperture and the acceptance angle of a fiber with a core index of 1.37 and a cladding index of 1.30.

2. If the fiber in Problem 1 has a core diameter of 50 μm, calculate the angles that would lead to light propagation.

3. Calculate the numerical aperture of skew rays propagating at an angle of 100° to the core radius in the fiber in Problem 1.

4. A fiber has 10 mW of light power injected into it. Of this, 9.3 mW exits the opposite end of the fiber. Determine the loss in dB.

5. If the light exiting the fiber in Problem 4 fell upon a detector with a light-sensitive area of 5 cm^2, determine the irradiance.

REFERENCES

Cherin, Allen H. *An Introduction to Optical Fibers.* New York: McGraw-Hill, 1983.

Senior, John M. *Optical Fiber Communications—Principles and Practice.* Upper Saddle River, NJ: Prentice Hall International, 1985.

Sterling, Donald J. *Technician's Guide to Fiber Optics.* Albany, NY: Delmar Publishers Inc., 1987.

Zanger, Henry, and Zanger, Cynthia. *Fiber Optics—Communications and Other Applications.* Upper Saddle River, NJ: Merrill/Prentice Hall, 1991.

4

TYPES OF OPTICAL FIBER AND THEIR PROPERTIES

Included in this chapter:

4.1 INTRODUCTION

Optical fiber used for communications may be one of several different designs. Advances in light sources, detectors, and manufacturing techniques have led to new designs that allow more data to be transmitted through longer distances. Fiber design has concentrated on reducing the losses in a fiber in two ways. Decreasing **attenuation** losses is focused on bringing as much of the light originally launched in the fiber out the other end. Reducing **dispersion** limits the amount of distortion in the signal carried by the light through the fiber.

The **loss** in optical fiber is defined differently depending on the type of loss described and the design of the fiber. Attenuation loss is generally measured in terms of the decibel (dB) which is a logarithmic unit. The decibel loss of optical power in a fiber is calculated through the formula

$$dB = -10 \log (P_{out}/P_{in}) \tag{4.1}$$

where

P_{out} is the power coming out of the fiber

P_{in} is the power launched into the fiber

The number of decibels lost is sometimes indicated by a negative sign to distinguish it from a gain light power which is represented by a positive number. However, many manufacturers assume that the negative sign for attenuation values is understood and do not include it.

 EXAMPLE 4.1

A signal of 100 mW is injected into a fiber. The outcoming signal from the other end is 40 mW. What is the loss in dBs?

To calculate loss in dBs

Using Equation 4.1,

$$dB = -10 \log(P_{out}/P_{in})$$
$$= -10 \log(40 \text{ mW}/100 \text{ mW})$$
$$= 3.98 \text{ dB}$$

Most fiber manufacturers characterize attenuation loss by the number of decibels lost per kilometer of fiber. This value can be calculated by

$$dB/km = -(10/L) \log(P_{out}/P_{in}) \tag{4.2}$$

where

P_{out} is the power coming out of the fiber
P_{in} is the power launched into the fiber
L is the length of fiber tested

The loss per kilometer (or dB/km) is a standard unit for describing attenuation loss in all fiber designs. However, some fiber-testing equipment will measure loss in the number of decibels based on an input power of 1 mW. This unit is abbreviated dBm. This calculation can be made using Equation 4.2 and substituting 1 mW for P_{in}.

EXAMPLE 4.2

A 2-km length of fiber has an input power of 20 mW and an output power of 150 μW. What is its loss in dB/km? Express this loss in dBms.

To calculate the loss in dB/km

Using Equation 4.2,

$$dB/km = -(10/L) \log(P_{out}/P_{in})$$
$$= -(10/2) \log(150 \text{ mW}/20 \text{ mW})$$
$$= 10.6 \text{ dB/km}$$

If a 1-mW signal was injected into the fiber, the loss would be 21.2 dB (10.6 dB/km × 2 km). Rearranging Equation 4.1 gives

$$P_{out} = (10^{-10.6/10}) \times 1 \text{ mW} = 0.09 \text{ mW}$$
$$dBm = -10 \log(0.09 \text{ mW}/1 \text{ mw})$$
$$= 10.5 \text{ dBm}$$

Dispersion loss can be described in one of two ways. The single-mode fiber design (see Section 4.4) expresses dispersion (Δt) as the difference between the width of a pulse of light launched into a fiber and the width of the same pulse as it exits the fiber. Dispersion is calculated by

$$\Delta t = (t_1{}^2 - t_2{}^2)^{1/2} \qquad \textbf{(4.3)}$$

where

t_1 is the width of the pulse at the output
t_2 is the width of the pulse at the input

 EXAMPLE 4.3

A pulse with a width of 10 ms is injected into a fiber. At the opposite end of the fiber, a pulse with a width of 12.5 ms emerges. Find the dispersion loss.

To calculate dispersion

Using Equation 4.3,

$$\Delta t = (t_1{}^2 - t_2{}^2)^{1/2}$$
$$= ((12.5 \text{ ms})^2 - (10 \text{ ms})^2)^{1/2}$$
$$= 6.5 \text{ ms}$$

For the multimode fiber design described in Section 4.4, dispersion is expressed as a **bandwidth length product (BWL)**. This value indicates the frequency of a signal that can be transmitted through a certain distance and is calculated by

$$\text{BWL} = 0.187/(\Delta t \ SW \ L) \tag{4.4}$$

where

SW is the spectral width of the source (see Chapter 5)
L is the length of the fiber tested

In many cases, fiber manufacturers will refer to the quantity calculated in Equation 4.4 as bandwidth rather than bandwidth length product. The length product can always be identified by the units kHz-km rather than just kHz which is used for bandwidth.

 EXAMPLE 4.4

A light source with a bandwidth of 100 GHz is injected into a 3-km fiber. If the pulse injected has a width of 20 ms and the bandwidth length product of the fiber is 100 kHz-km, what is the pulse width at the exit end of the fiber?

Equation 4.4 is used. Rearranging the equation by solving for Δt gives

$$\text{BWL} = 0.187/(\Delta t \ SW \ L)$$
$$\Delta t = 0.187/(\text{BWL} \ SW \ L)$$
$$= 0.187/(100 \text{ kHz-km} \times 100 \text{ GHz} \times 3 \text{ km})$$
$$= 6.2 \text{ ps}$$

The pulse will, therefore, change its width by 6.2 ps and will be very nearly the same size.

Example 4.5 illustrates the calculations used for the quantities dB, dB/km dispersion, and bandwidth length product.

 EXAMPLE 4.5

A fiber is tested for attenuation and dispersion losses. If the input power for the fiber is 3 mW, the output power is 1.5 mW, and the fiber is 10 km long, calculate the dB attenuation and attenuation per kilometer.

Using Equation 4.1

$$dB = -10 \log(P_{out}/P_{in})$$
$$= -10 \log(1.5/3)$$
$$= 3.0 \text{ dB}$$

Using Equation 4.2

$$dB = -(10/L) \log(P_{out}/P_{in})$$
$$= -(10/10) \log(3/1.5)$$
$$= -3 \text{ dB/km}$$

If a pulse with a width of 0.20 ms is sent from a light source with a spectral width of 10 nm through the same fiber and exits with a width of 0.32 ms, calculate the dispersion and the bandwidth length product.

Using Equation 4.3

$$\Delta t = (t_1^2 - t_2^2)^{1/2}$$
$$= (0.32^2 - 0.20^2)$$
$$= 0.25 \text{ ms}$$

Using Equation 4.4 and converting nm and km to meters,

$$BWL = 0.187/(\Delta t\ SW\ L)$$
$$= 0.187/[(10 \times 10^{-6})(10 \times 10^3)]$$
$$= 1.87 \text{ Hz-km}$$

In this chapter, the various mechanisms that cause both types of losses are discussed. Designs that reduce one or the other or both are then described, and the types of fiber used for different applications are identified.

4.2 ATTENUATION LOSS IN FIBER

The loss of light in a fiber occurs because of two main mechanisms. As light passes through the fiber, the material in the fiber causes **absorption** of a percentage of it. At the same time, impurities and imperfections in the fiber create **scattering** of some of the light. Loss of light also occurs when there are small bumps or variations in the surface of the core of the fiber. These variations are known as **microbends**. Light is lost when it is first launched into the fiber and at any splices or connections along the fiber length (discussed further in Chapter 6).

Absorption of light by the fiber depends greatly on the wavelength of light used and on any impurities in the fiber itself. As shown in Figure 4.1, the amount of light absorbed generally decreases with an increase in wavelength. At roughly 850 nm, the absorption is 1.5 dB/km, whereas at 1500 nm, it is less than 0.5 dB/km. However, there are three bands of wavelengths where the absorption increases drastically. These bands, centered around 950 nm, 1250 nm, and 1380 nm, are caused by moisture in the fiber. Hydroxyl ions (OH^-) in the moisture absorb the light in these wavelength bands.

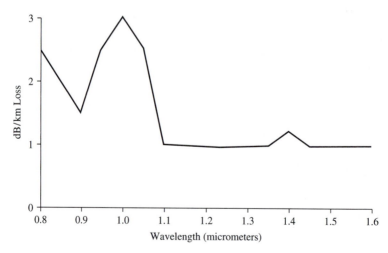

Figure 4.1 Typical Material Absorption Loss in Fiber

Scattering of light in a fiber is also wavelength dependent. Changes in the index of refraction of the fiber that are caused by microscopic variations in the concentration of the glass lead to an effect known as **Rayleigh scattering**. This type of scattering is inversely proportional to the fourth power of the wavelength or,

$$\text{loss } \alpha \; 1/\lambda^4 \tag{4.5}$$

Longer wavelengths of light scatter less than shorter wavelengths.

Microbend loss in a fiber occurs when the core surface has small variations in shape. These variations, or bumps, change the angle at which light strikes the core to cladding interface and can cause the light to refract into the cladding rather than reflect into the core. Microbends can be caused during the manufacturing of the fiber or may occur during the handling of the fiber. Variations caused by handling can be greatly reduced by protective cable designs used in commercial grade communications fiber.

4.3 DISPERSION LOSSES

When data is sent along communications fiber, information is generally contained in pulses in the intensity of the light (see Chapter 7 for more details). These pulses can become distorted as they travel through the fiber. Three mechanisms cause this distortion in the signal. **Modal dispersion** occurs in fibers that have more than one mode (see Chapter 3). **Material dispersion** is a wavelength-based effect caused by the glass of which the fiber is made. **Waveguide dispersion** occurs only in fibers with a single mode.

Modal dispersion results because the various modes in the fiber core are different lengths. Recall from Chapter 3 that as light travels through the fiber, it spreads to fill all the possible modes or paths through the core. Since each path is a different length, some of the light reaches the end of the fiber sooner than the rest. The result is the pulse of light containing the data transmitted is elongated or stretched out. If the pulses stretch too much, they begin to overlap each other and may reach a point where one is not distinguishable from the next. Signals that undergo too much modal dispersion become garbled at the other end of the fiber.

The difference in the time it takes light in a longer mode to reach the end of the fiber versus the time it takes light in a shorter mode can be calculated using the approximation

$$\Delta t/L = (n_{\text{core}})^2(1 - \pi/V)L/c \tag{4.6}$$

where

> n_{core} is the index of refraction of the core
> V is the V number of the fiber (see Chapter 3)
> L is the length of the fiber
> c is the speed of light (3×10^8 m/s)

and the resulting number is in seconds per length of fiber in meters. Example 4.6 demonstrates how to calculate the differences for a common fiber.

 EXAMPLE 4.6

Calculate the difference in time per meter between a long mode and a short mode in a fiber with a V number of 23 and a core index of 1.46.

Using Equation 4.5

$$\Delta\tau/L = (n_{core})^2(1 - \pi/V)L/c$$
$$= 1.46^2(1 - 3.14159/23)10/3 \times 10^8$$
$$= 61.3 \text{ ns/km}$$

Like modal dispersion, material dispersion causes the pulses in the signal to stretch out. In the case of material dispersion, however, the change in size is because different wavelengths of light travel at different speeds in the glass that makes up the fiber core.

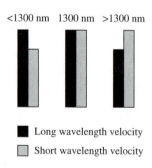

Figure 4.2 Material Dispersion

At shorter wavelengths (less than 1300 nm), the long wavelengths travel faster than the shorter wavelengths (see Figure 4.2). In other words, an 855-nm light wave would reach the end of the fiber quicker than an 840-nm light wave. Above 1300-nm, the effect is reversed. A 1500-nm wave would travel faster than a 1550-nm wave. The key to modal dispersion is the 1300-nm wavelength. At this length, the difference in speeds is minimized so that even though a 1305-nm wave would be faster than a 1295 nm, the difference in speeds would be slight. For this reason, 1300 nm is often called the **zero dispersion wavelength**.

Waveguide losses occur in fiber that is designed so that it has only a single mode or path for the light to follow. Because such a design has a very small core, some of the light actually travels in cladding of the fiber. The cladding has a different index of refraction (see Chapter 3), and light traveling through it will reach the end of the fiber sooner than light traveling in the core.

Initially because of available light sources and then later as an effort to improve fiber characteristics, the standard operating wavelength for communications fiber has increased from 870 nm to 1300 nm and finally to 1550 nm. This increase in wavelength has led to a general decrease in absorption and dispersion.

4.4 TYPES OF FIBER

Different designs in optical fiber have been produced to reduce the amount of loss. Dispersion loss, in particular, has been a driving force in fiber design. Table 4.1 summarizes the types of fiber designs and the problems they were intended to correct.

The **index of refraction profile** of a fiber design is one of the strong factors in determining the amount of modal dispersion in a fiber. The original fiber design described in Chapters 1, 2, and 3 is a **step index fiber** in which the core has one index of refraction and the cladding has another. A diagram of the index of refraction of a cross sec-

Table 4.1 Types of Optical Fiber

Fiber Type	Corrected Loss
Step Index Multimode	Original Design
Grade Index Multimode	Modal Dispersion
Single mode	Modal Dispersion

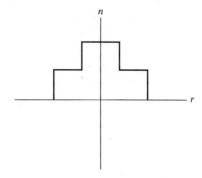

Figure 4.3 Step Index Fiber

tion of the fiber demonstrates this profile. As shown in Figure 4.3, the step index fiber has a profile that resembles a step on either side.

To reduce the amount of modal dispersion in the original fiber design, **graded index fiber** has a step index profile with a gradual change in index from core to cladding. The resulting fiber allows light in longer modes to travel faster than light in shorter modes and reduces the modal dispersion of the fiber. The index of refraction profile for a graded index fiber is shown in Figure 4.4.

The single-mode fiber is designed to eliminate modal dispersion by reducing the number of modes in the fiber to one. Recall from Chapter 3 that the number of modes depends on the size of the core in a fiber. Single-mode fiber has a very small core diameter so that only one mode exists. The result is zero modal dispersion, but the trade-off is in the

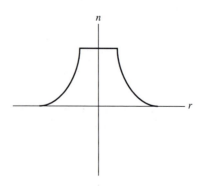

Figure 4.4 Graded Index Fiber

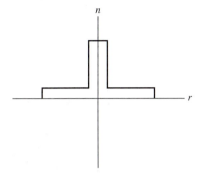

Figure 4.5 Single-mode Fiber

addition of waveguide dispersion as discussed in Section 4.3. Although waveguide dispersion is relatively nominal, single-mode fiber does require a much more sophisticated light source in order to launch enough light into the tiny core. This means the fiber has an increase in installation and operating costs that outweigh the advantages for most short-distance fiber applications. However, long-distance telephone lines and other long-distance applications use single-mode fiber exclusively because of its advantages.

The index of refraction profile for a standard single-mode fiber is shown in Figure 4.5. Notice that this design incorporates the step index profile since graded index is no longer necessary for reducing modal dispersion. Other single-mode fibers incorporate parabolic or triangular profiles which allow for single-mode operation at larger diameters.

Another fiber design uses different types of glass to create a **dispersion shifted fiber** which allows the zero dispersion wavelength to occur at 1550 nm rather than 1300 nm. Some fibers are also made of plastic or a plastic cladding with a glass core known as **plastic clad silica** (**PCS**). The plastic fibers are not normally used in communications applications because of their high losses.

4.5 INSTALLATION AND HANDLING CONSIDERATIONS IN FIBER TYPES

The technician dealing with optical fiber of different types on a daily basis will begin to see some common rules of thumb involving identification, handling, and applications. In general, a distinction between single-mode and multimode fiber is the easiest comparison.

Single-mode has, by its very nature, a much smaller core diameter than multimode. However, the cladding diameter is normally such that the overall size of the fiber appears to be the same. You can quickly determine the type of fiber by finding the core/cladding diameter designation of the fiber. Most fibers are labeled with a number such as 50/100 or 10/100. The first number is the core diameter, and the second is the cladding diameter—both expressed in microns. Multimode fiber normally has a larger core diameter (usually on the order of 50 microns). To illustrate core and cladding diameters (and shapes), shine a visible light laser beam (or a well-focused LED) into a short piece of fiber. By placing an expanding lens and a white piece of paper near the output end of the fiber, a general picture of core and cladding is produced.

Because its core is so much smaller, the single-mode fiber requires specialized connectors, light sources, and detectors. Some of these considerations are discussed later in this book. These specialized components mean increased cost, making single-mode fiber a more expensive choice and not always the most viable one, depending on system requirements. Short distances and low frequencies can normally be accommodated by multimode fiber.

CHAPTER REVIEW

New Terms

Section 4.1
Losses
Attenuation
Dispersion
Decibel
Bandwidth Length Product

Section 4.2
Absorption
Scattering

Section 4.3
Modal Dispersion
Material Dispersion
Waveguide Dispersion
Zero Dispersion Wavelength

Section 4.4
  Index of Refraction Profile
  Step Index Fiber
  Graded Index Fiber
  Single Mode
  Plastic Clad Silica
  Dispersion Shifted Fiber

Other Chapters with Related Information

  Chapter 3
  Chapter 7
  Chapter 8
  Chapter 10

Review Questions

1. What are the two types of loss in optical fiber?
2. List three causes of distortion. Suggest methods for reducing their effects.
3. In what applications would you expect to use single-mode fiber?
4. Sketch the index of refraction profile for triangular and parabolic single-mode fibers.
5. What kind of attenuation would you expect at 910 nm?
6. If you were asked to choose a fiber to use in a local area network that requires 10 meters of cable, which type would you choose? Why?
7. Explain the difference between dispersion and bandwidth length product.

Thought Questions

1. Describe the problems that may be associated with using a fiber to transmit light with a wavelength of 5000 nm.
2. Why wouldn't you use single-mode fiber?
3. What types of fiber are most likely to be used in a long-distance telephone connection? A local area network? The information superhighway?
4. How would making a fiber out of plastic affect its performance?
5. What must be considered to minimixe loss when manufacturing a fiber?

Problems

1. Calculate the dB loss of the following fibers if the input power is 10 mW and the output powers are the values given.

   (a) 9 mW   (b) 7 mW   (c) 5 mW   (d) 3 mW

2. Using the output powers listed in Problem 1, calculate the dBm for each fiber.

3. If a fiber had an attenuation of 3 dB in a length of 1 km, what would be its loss per kilometer?

4. Calculate the dispersion of a fiber with an incoming pulse length of 5 ms and an outgoing length of 5.5 ms.

5. Calculate the bandwidth length product of a fiber with a dispersion of 0.1 ms in a length of 3 km.

6. What is the difference in mode travel times for a fiber with a core index of 1.5 and a V number of 120?

7. A fiber manufacturer says its brand of fiber has a bandwidth length product of 400 kHz-km. Could you send a 800-kHz signal through this fiber over a distance of 0.3 km?

REFERENCES

Hecht, Jeff. *The Laser Guidebook,* 2nd ed., Blue Ridge Summit, PA: TAB Books, 1992.

Senoir, John M. *Optical Fiber Communications—Principles and Practice.* Upper Saddle River, NJ: Prentice Hall International, 1985.

Sterling, Donald J. *Technician's Guide to Fiber Optics.* Albany, NY: Delmar Publishers, Inc., 1987.

Sze, S. M. *Semiconductor Devices Physics and Technology.* New York: John Wiley & Sons, 1985.

Zanger, Henry, and Zanger, Cynthia. *Fiber Optics—Communications and Other Applications.* Upper Saddle River, NJ: Merrill/Prentice Hall, 1991.

5

LIGHT SOURCES FOR OPTICAL FIBERS

Included in this chapter:

5.1 INTRODUCTION

Two types of light sources are commonly used for optical fiber in communications applications. These sources are the **light-emitting diode (LED)** and the **semiconductor laser** (or laser diode). These two sources have distinct characteristics in terms of performance, cost, and ease of use. The choice of light sources for a particular application depends on these characteristics. The selection is usually based on the higher cost and higher performance of the laser versus the lower cost and lower performance of the LED. The performance benefits of the laser outweigh its costs for long-distance or high-volume communications although LEDs have been developed that can perform nearly as well in some instances.

This chapter describes these two light sources in detail, including the theory behind their operation, their performance characteristics, and standard methods for setup and use. The information about theory of operation is rooted in semiconductor physics; therefore, a quick review of this topic may be useful. Students who are unfamiliar or out of practice with the concepts involved should consult one of the excellent references listed at the end of the chapter.

This chapter also covers the role of the source in the entire communication transmitter. Again, the principles involved are rooted in another field (this time, solid-state electronics), and a review of one of the references may be helpful.

5.2 LIGHT-EMITTING DIODES

The light-emitting diode has a wide range of applications as a source of light. In fiber optics, the LED offers low price, ease of use, and a well-established technology. It consumes only a minimal amount of electrical power and doesn't require any specialized devices to operate.

The LED is a diode, and, like any diode, it is based on the construction of a layer of semiconductor materials. A semiconductor is a material that consists of a regular pattern of atoms or molecules locked together in a crystal lattice. The atoms in the lattice are held together by a **covalent bond**, or a sharing of electrons. Each atom shares electrons with other atoms in order to obtain eight electrons in the outermost (or **valence**) **orbital**. A simplified diagram of a crystal lattice structure is shown in Figure 5.1.

To produce a semiconductor, in a structure such as the one shown in Figure 5.1, some of the atoms are replaced with atoms of another type in a process called **doping**. The replacement atoms may require either one more or one fewer electron to complete their valence orbital. As a result, when the replacement atom requires an additional electron, there is an empty slot or **hole** that an electron could fill (the other atoms in the lattice do not supply the missing electron). Such a material is called a **p-type** semiconductor. If the replacement atom has an extra electron, that electron is not shared with any other atom, and the resulting material is known as an **n-type** semiconductor.

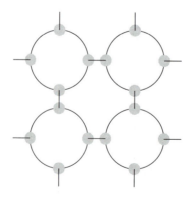

Figure 5.1 Crystal Lattice Structure

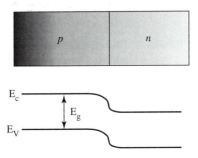

Figure 5.2 p-n Junction.

To produce an LED (or any diode), a piece of n-type semiconductor is sandwiched together with a piece of p-type semiconductor to form a **p-n junction**. This junction has special electrical characteristics due to its structure.

Consider the p-n junction shown in Figure 5.2. At the edge between the two types of materials, extra electrons from the n-type material cross over to fill the holes in the p-type in a process known as **recombination**. The result is an area at the intersection between the materials where there are no holes or free electrons. This area, known as the **depletion region**, acts as a barrier to the rest of the electrons in the n-type so that they are unable to cross over and fill holes in the p-type. This process is completed as soon as the junction is formed. The resulting device is then connected to a source of electricity to produce light or to perform other functions in a circuit.

If, for example, the positive electrode of a DC source of electricity is connected to the n-type material and the negative electrode is connected to the p-type, the junction is said to be **reverse biased**. Under these conditions, electrons from the source may fill some holes in the p-type material, but, because of the depletion region, they cannot make a complete pass through the circuit. The result is that there is no flow of electricity.

If the junction is **forward biased**, the positive electrode is connected to the p-type material, and the negative electrode is connected to the n-type. Electrons are pulled from the p-type material and injected into the n-type. The electrons in the n-type move across the depletion region to fill the spaces left behind in the p-type by those electrons that are drawn away by the source. Current flows through the circuit, and, each time an electron "drops" into a hole, it releases energy (see Figure 5.3). The energy that is released is some form of electromagnetic radiation or heat depending on the selection of materials used in producing the semiconductor. The frequency (or wavelength) of the radiation also depends on the atoms involved and their respective energy structures.

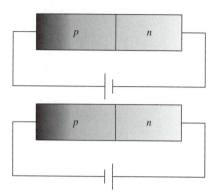

Figure 5.3 Forward and Reverse Bias

Each type of atom has a unique number of electrons that are arranged in a well-defined pattern around the nucleus.[1] The arrangement of the electrons results in a unique amount of energy for that atom. When the electrons are removed or added to the atom, the result is absorption or emission of energy by the atom. Those atoms that lose an electron have absorbed enough energy to pull the electron away from the atom, and those atoms that gain an electron release energy equal to the amount needed to remove an electron.

A useful method of visualizing the gain or loss of energy by an atom is to consider the electrons as part of the **valence band** or part of the **conduction band** of an atom. The valence band holds those electrons that are still attached to the atom, and the conduction band holds the electrons that have been freed from the atom. In Figure 5.3, the valence bands and conduction bands for the two materials in a p-n junction are shown. Notice that the n-type material, because of its free electrons, has a conduction band that is at a much lower energy than that of the p-type. When electrons in a forward biased p-n junction pass from the conduction band of the n-type to the valence band of the p-type, they release energy equal to the difference in energy between the n-type conduction band and the p-type valence band (known as the **band gap energy**).

---

[1] *Note*: Even though the atoms are arranged in the atom by a set of rules defined by their quantum numbers, their exact location at any given time is not known. This concept is explained thoroughly in any text dealing with modern physics and quantum mechanics. For more information, please consult one of the references listed at the end of this chapter.

The result of all this energy information is revealed by returning to some of the principles discussed in Chapter 2. Remember from the discussion on quantum optics, the wavelength or frequency of light (or any electromagnetic radiation, for that matter) is related to the energy of its photons. In the case of the p-n junction, the photon energy is a direct result of the types of atoms used to produce the semiconductor material.

To produce light-emitting diodes, semiconductor materials are formed from atoms with three valence electrons to share and atoms with five valence electrons to share (known as **Group III** and **Group V** atoms, respectively, because of their position in the periodic table). If these two types of atoms were combined in equal amounts in a lattice, the result would be a material with no free electrons and no holes. However, by adjusting the ratios of Group III and Group V atoms, both n-type and p-type materials can be formed.

To produce light in the 800-nm region, LEDs are constructed from the Group III atoms gallium (Ga) and aluminum (Al) and the Group V atom arsenide (As). The resulting junction is commonly abbreviated GaAlAs for gallium aluminum arsenide. For longer wavelengths, gallium is mixed with the Group III atom indium (In) and arsenide with the Group V atom phosphate (P) to form a gallium indium arsenide phosphate (GaInAsP) junction. Other materials are often used to produce different wavelengths of light. Table 5.1 lists some common semiconductor materials and their resulting output wavelengths.

A p-n junction made from two different mixtures of the same types of atoms (known as a **homojunction**) does not produce a very useful light source for an optical fiber. Light from this type of junction is emitted in all directions, and only a small amount of the total light produced is directed at the fiber. The efficiency of the output (conversion of electricity to light) is also very low.

To alleviate the loss of light in a homojunction device, LEDs for use in optical fiber are made from a p-type material of one set of atoms and

Table 5.1 Semiconductor Materials and Resulting Wavelengths

Material	Wavelength (nm)
AlGaInP	630–680
GaInP	670
GaAlAs	620–895
GaAs	904
InGaAs	980
InGaAsP	1100–1650
InGaAsSb	1700–4400

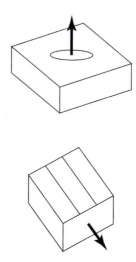

Figure 5.4 Edge Emitters and Surface Emitters

an n-type from another set. The resulting **heterojunction** device confines the **charge carriers** (electrons and holes) and the light to a smaller area. The light is then emitted from the "edge" of the sandwich of materials. Because of the difference in emission patterns, homojunction devices are often known as **surface emitters** and heterojunction devices as **edge emitters**. The difference in emissions is illustrated in Figure 5.4.

The materials in a heterojunction device are layered to enhance the concentration effect. As shown in Figure 5.4, the n-type material is generally two layers, as is the p-type. The entire junction is usually manufactured on a sort of backing or *substrate* of material and then sandwiched between metal contacts used to hook the device to a source of electricity.

5.3 SEMICONDUCTOR LASERS

The word *laser* is an acronym for <u>L</u>ight <u>A</u>mplification by <u>St</u>imulated <u>E</u>mission of <u>R</u>adiation. In general, lasers differ from other light sources because they produce light that is highly directional, coherent, and **monochromatic** (all one wavelength). The process by which a laser produces this unique light is rooted in the mechanisms behind light production.

In Chapter 2, the production of light was described in terms of an atom. When an atom absorbs energy, its electrons are affected. In the previous section, we examined the situation in which an electron is completely removed from an atom by this absorption of energy. However, many atoms do not lose their electrons when they absorb energy. The electrons are only repositioned. When this happens, the affected electron will eventually return to its original position, and the atom will release the energy that it absorbed earlier. The released energy is often some form of electromagnetic radiation. The process by which it is released by the atom is known as **spontaneous emission**. The emission is spontaneous because the atom will release the energy of its own accord.

Another possible scenario for atom-releasing energy is illustrated in Figure 5.5. If an atom that has absorbed energy and had an electron repositioned collides into a photon released by some other atom, the first atom will release its energy as a result of the collision. In this case, the energy release is known as **stimulated emission** since it was caused by the interaction with the photon.

Stimulated emission is the key to the operation of a laser. When a photon is produced by stimulated emission, it is coherent with the photon that collided with the atom. In other words, the two photons are *in-phase* (see Chapter 2 for more details). Furthermore, because the process only occurs with two photons that have the same energy (and, therefore, the same wavelength), the photons are monochromatic. Two of the properties of laser light are the result of stimulated emission.

The third property of laser light—directionality—is the result of the process used to produce stimulated emission. To produce stimulated emission, a group of atoms must be injected with energy and then collide with photons. This situation is easy to reproduce since any group of atoms that are injected with energy will undergo spontaneous emission. The resulting photons can then collide with atoms that have not yet released their energy and cause stimulated emission.

This process happens in every source of light. Unfortunately, the amount of spontaneous emission generally outweighs the amount of

Figure 5.5 Stimulated Emission

stimulated emission so that the coherent and monochromatic photons are overwhelmed by ordinary light. To increase the amount of stimulated emissions, the majority of the atoms in a light source must have extra energy. If this is the case, any photon released via spontaneous emission will have a better chance of colliding with an atom with extra energy rather than with an atom that has already released its photon. The group of atoms that have more atoms with extra energy than without is said to have a **population inversion**. This situation can only occur when a large amount of energy is pumped into the atoms and when the atoms retain that energy for a relatively long period of time.

The other mechanism used to ensure stimulated emission involves optical feedback. If the light source has a reflective surface at either end, the photons traveling toward these surfaces will be reflected into the source. Each time the photons pass back through the atoms, they can produce more stimulated emission. As long as the atoms are continuously pumped with energy, they will be available to collide and produce additional stimulated photons. The added bonus is that all the light traveling between the reflective surfaces points in the same direction.

To produce a semiconductor laser, an LED design is modified to include reflecting surfaces at each end. These surfaces are created by **cleaving** the sides of the LED. Cleaving is a process that takes advantage of the crystal structure of the semiconductor materials. Just as a diamond can be cut along certain lines to produce a flat surface and a piece of wood can be split along the grain, semiconductor materials can be cut at the proper angle to form flat surfaces. These surfaces then act as reflectors for the laser. These surfaces only reflect a portion of the light (on the order of 5%) and let the rest pass through. The light that passes through forms the beam of the laser.

Like the LED (see Section 5.1), the semiconductor laser takes advantage of a heterojunction structure to confine the light and the charge carriers to a smaller area. This concentration of light and electricity improves the efficiency of the laser and allows operation at room temperature. Currently, the double heterojunction design is used to produce semiconductor lasers. This design uses the p-n junction sandwiched between two layers of a different material (a GaAs p-n junction sandwiched between two layers of GaAlAs on each side, for example). The two layers confine the light to a smaller area because of their higher index of refraction. To further confine the active part of the laser (and increase current density), the active layer is applied as a stripe between 1 to 10 μm wide rather than a layer that is the entire width of the device. Figure 5.6 illustrates a common double heterojunction laser.

The stripe of active layer can be produced using several methods including a buried channel formed by etching out all but the narrow

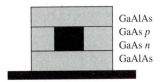

GaAlAs
GaAs *p*
GaAs *n*
GaAlAs

Figure 5.6 Double Heterojunction Laser

stripe, a channel formed into the surrounding layer which is then filled by growing the active layer into it, or two channels or a channel that is overfilled or underfilled. One of the most interesting concepts comes when the layer is so thin that the quantum mechanical properties of the material affect the energy levels involved. These **quantum well** designs can lead to much lower operating currents and higher output powers.

The condition of population inversion in a laser gives rise to an operating parameter of a semiconductor laser known as **threshold current** (or *threshold current density*). A laser beam is not produced until the current in the circuit reaches a minimum level. After this threshold is reached, the output light power of the laser increases rapidly with current until a maximum output is reached. Figure 5.7 illustrates the concept of threshold current.

Like LEDs, semiconductor lasers rely on a combination of Group III and Group V materials to produce the necessary p-n junctions. Generally, a combination of aluminum, gallium, and antimony (Sb) or a combination of gallium, indium, arsenide, and phosphate is used. Wavelengths in the red portion of the visible spectrum (about 700 nm) and the infrared are possible.

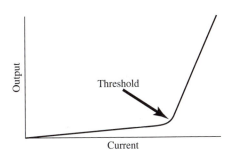

Figure 5.7 Threshold Current in a Semiconductor Laser

5.4 OPERATIONAL PARAMETERS

The LED and the semiconductor laser are characterized as light sources for optical fibers by certain operational parameters. These parameters describe the electrical and optical characteristics of the sources as well as their performance in terms of **modulation**. This last area covers the source's ability to represent data in the form of pulses of light and is most concerned with the speed at which pulses can be produced. The following brief definitions are designed to provide a background adequate for choosing the proper light source for the job. The end of this section describes the advantages of the LED and the semiconductor laser.

Output Power. The amount of light produced by a light source is generally characterized by the optical power. Normally expressed in milliwatts or microwatts, the output power of a source is one factor involved in the distance that a signal may be sent along an optical fiber.

Spectral Width. As described in Chapter 4, the type and number of wavelengths emitted by an optical source can be a major factor in the amount of loss in a fiber. Light sources are described in terms of their **spectral width**. Spectral width data is often provided in graphical form. The graph illustrates the optical power output for each wavelength emitted by a light source. Figure 5.8 is an example of such a graph.

The spectral width can also be a numerical value which represents the width of the graph at half of the maximum power (the so-called **full width at half maximum** or **FWHM**) and is expressed in nanometers,

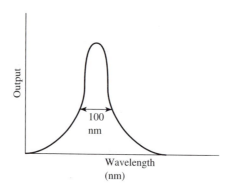

Figure 5.8 Spectral Width of a Typical LED

micrometers, or angstroms. Many source manufacturers provide both the number and the graph.

Rise Time (or Speed of Operation). The speed at which a light source can be modulated (or turned off and on to produce pulses) can be described in terms of the **rise time**. When a light source is turned on to produce a pulse of light, it takes a measurable amount of time to produce the maximum output power. As illustrated in Figure 5.9, the time the source takes is normally measured between the point at which the source is at 10% of maximum power and the point at which it is at 90%. The rise time (t_r) can be used to find the bandwidth of operation for the source (BW) by the equation

$$BW = 0.035/t_r \qquad (5.1)$$

Sometimes bandwidth is used to describe the speed of operation directly, and sometimes the rise time is given.

In some cases, the **fall time** of the source is also given. This number provides the time it takes a source to turn off and is measured between the 90% power and 10% power points at the end of the pulse (see Figure 5.9).

Numerical Aperture. Just as a fiber has a numerical aperture (see Chapter 3) that describes the physical area that light must enter in order to propagate through the fiber, light sources have a numerical aperture that describes the size of the output pattern of the light. As with fibers, numerical apertures for light sources can be calculated in terms of the angle involved with the output. As shown in Figure 5.10, the output pattern can be described by the angles $\theta_\perp$ and θ_p which measure the spread of the light in both vertical and horizontal directions.

Figure 5.9 Rise Time and Fall Time

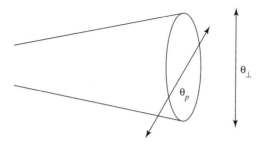

Figure 5.10 Output Patterns of Light

Cost. As with any device used in an application, cost is a factor for optical sources in optical fibers. The cost issue tends to be a trade-off between performance and price, and the specific application and its properties is the deciding factor.

Ease of Operation. The ease of operation varies greatly between the LED and the semiconductor laser. The LED generally needs a much less sophisticated power supply, is more sturdy, and is able to operate over a wider range of environmental conditions. The laser requires a more complex source of electricity and is more susceptible to temperature fluctuations.

Forward or Supply Voltage. The amount of voltage required to produce light from a source can be described in terms of a maximum and minimum. The minimum voltage is the amount needed to provide adequate light output, and the maximum is the upper limit beyond which damage to the source can occur.

Operating and Threshold Current. The amount of current to operate the light source is, of course, a function of the supply voltage and the circuit resistance. As described in Section 5.5, most light sources require resistors in series to limit the current to a level that will not damage the source. Operating currents are optimum values for light output. As mentioned in Section 5.4, lasers require a minimum, or threshold, current in order to produce a beam.

Packaging. Light sources can be packaged a number of different ways. For optical fibers, the light sources normally include a tiny **microlens** for focusing the light and reducing the numerical aperture. Some sources include a **pigtail** or short section of fiber affixed to the surface of the source or positioned very close to the surface to allow for

Table 5.2 Advantages of LEDs and Semiconductor Lasers

Characteristic	LED	Laser
Output Power	Lower	Higher
Spectral Width	Wider	Narrower
Numerical Aperture	Larger	Smaller
Rise Time	Slower	Faster
Cost	Less	More
Ease of Operation	Easier	More Complicated

easier coupling to the fiber. Light sources may also have standard fiber optic couplers (see Chapter 7 for more information) as part of their packaging.

As a summary of operational characteristics, Table 5.2 illustrates the advantages of LEDs and semiconductor lasers for each of the characteristics discussed. Although a laser offers several operational advantages, the LED is less expensive and easier to use. For this reason, the LED is normally used for short- and medium-distance communications, and the laser is confined to long-distnce and single-mode fiber applications.

 EXAMPLE 5.1

The data for two LED sources are given. If the sources are to be used in a fiber optic system that requires the following operational parameters, which LED would be the best choice (ignore for the moment the effects of splices and connectors and detector properties)?

System Requirements

Bandwidth = 20 MHz

Output power of 10 μW after 1.5 km (loss of 3 dB/km)

Electrical power maximum of 1 watt

Fiber Numerical Aperture = 1.2

LED Characteristics

Parameter	LED #1	LED #2
Output power	1 mW	2 mW

Rise Time	1 ns	10 ns
NA	1.10	1.70
Supply Voltage	2.3	3.0
Forward Current	10 mA	50 mA

(All values are typical. Most data would be given in ranges rather than single values.)

To determine the bandwidth for both LEDs

Using Equation 5.1,

$$BW = 0.035/t_r$$

LED #1

$$BW = 0.035/1 \text{ ns} = 35 \text{ MHz}$$

LED #2

$$BW = 0.035/10 \text{ ns} = 3.5 \text{ MHz}$$

Electrical power consumption comes from the familiar, IV equation in electronics.

LED #1

$$P = IV = 23 \text{ mW}$$

LED #2

$$P = IV = 150 \text{ mW}$$

Output power comes from Equation 4.1.

$$dB = -10 \log(P_{out}/P_{in})$$

so P_{out} is equal to

$$P_{out} = 10^{-(dB/10)}P_{in}$$

where the dB loss is 3 dB/km $\times$ 1.5 km = 4.5 dB. The output power for the two LEDs would then be

$$P_{out} = 10^{-(4.5/10)}(1 \text{ mW}) = 0.36$$

LED #2

$$P_{out} = 10^{-(4.5/10)}(2 \text{ mW}) = 0.71$$

LED #1 has the advantage in bandwidth and meets all the requirements. LED #2 does not have a high enough bandwidth.

5.5 OPERATIONAL SETUPS

The systems used to operate light sources for optical fibers depend on the characteristics of the source and the need for fast, clean modulation. For both the LED and the semiconductor laser, the electrical source provides a modulated DC voltage and contains current limiting resistors. Modulation can be accomplished through the use of transistors, digital logic gates, or operational amplifiers depending on the desired results. The semiconductor laser requires additional drive circuitry for producing a steady output.

To produce light from an LED or a semiconductor laser, all that is required is a DC voltage source and a resistor large enough to maintain the current at the operational value. The source could be something as simple as a combination of AA batteries. Unfortunately, the light produced with such a setup would be of little use to a modern fiber optic communications system. The light output would have a tendency to fluctuate (by large amounts in the case of a laser), and there is no system for modulation.

Typical electrical supplies (or *drivers*, as they are called) for optical sources in fiber have a well-regulated output and a modulating system. The regulation system for an LED can take advantage of technology invented to regulate the supplies of integrated circuits used so extensively in electronics. A standard voltage regulator maintains the output voltage at a constant level.

For a semiconductor laser, the susceptibility of the laser to two main fluctuations must be considered. The output of the semiconductor laser varies widely with both temperature and current. As the current increases in a semiconductor laser, its output power increases and its spectral output shifts. To exaggerate the point, a laser producing 10 mW at 700 nm at one current level may produce 12 mW at 710 nm at another. Such shifts in output can wreak havoc with a communications system running at maximum speed.

In addition to current, temperature affects the output of the laser as well. The semiconductor laser will drop in output power when the temperature increases and will also shift its wavelengths. This second problem can be even more difficult to control than current fluctuations.

Several design additions have been provided to maintain a steady output from the laser. Current regulation in the driver is a first step to providing a constant operating current. Heat sinks, heat sink compound, and proper packaging contribute to maintaining a constant temperature.

To enhance the control of its output, semiconductor lasers are equipped with a **monitor photodiode**. The monitor photodiode measures the output power of the laser and provides a signal back to the driver circuitry. Fluctuations in output power are compensated for by adjustments in the supply voltage (and, subsequently, the current in the laser). The monitor photodiode is normally packaged with the laser and appears as a third lead extending from the housing of the laser.

To modulate the output of an optical source, it can be wired in a circuit like that shown in Figure 5.11. The source (in this case, an LED) is connected to the emitter of a standard NPN transistor. If the voltage to the base of the transistor reaches the saturation point, the current to the LED equals the saturation current which is chosen to be equal to the operation current for the LED. The size of the resistor and the type of transistor control the parameters of the circuit. Pulses sent to the base of the transistor appear as pulses of light from the LED.

The optical source can also be modulated using a standard logic gate circuit. As shown in Figure 5.12, the output of an AND gate is tied to the base of a transistor. One input of the AND gate is "tied high" or kept at a voltage that corresponds to a 1 on that input. The other input acts as the modulating signal since if it is 1, the base is saturated and

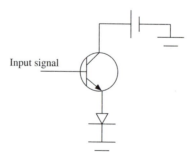

Figure 5.11 Transistor Modulation

Figure 5.12 Logic Gate Modulation

the LED turns on. If the second input is a zero, the base is tied to ground, and the LED is turned off. Similar results can be obtained using a NAND gate or some other combination of logic gates.

To produce an analog output from a light source, an operational amplifier (op amp) can be used. The op amp may drive the source directly or as part of its feedback loop. The voltage to the positive (+) input of the op amp is varied to produce varying amounts of optical output power.

CHAPTER REVIEW

New Terms

Section 5.1
   Light-emitting Diode (LED)
   Semiconductor Laser

Section 5.2

Covalent Bond	Conduction Band
Valence	Band Gap Energy
Doping	Group III
Hole	Group V
P-type	Homojunction
N-type	Heterojunction
P-N Junction	Charge Carrier
Depletion Region	Surface Emitter
Reverse Bias	Edge Emitter
Forward Bias	Substrate
Valence Band	

Section 5.3
   Light Amplification by Stimulated Emission of Radiation
   Monochromatic
   Spontaneous Emission
   Stimulated Emission
   Population Inversion
   Cleaving
   Threshold Current

Section 5.4
   Modulation
   Output Power
   Spectral Width
   Full Width at Half Maximum
   Rise Time
   Fall Time
   Microlens
   Pigtail

Section 5.5
   Driver
   Monitor Photodiode

Review Questions

1. Discuss the advantages and disadvantages of the two available optical sources for fiber optics.
2. Describe the two types of semiconductor materials.
3. What happens when a p-n junction is forward biased? Reversed biased?
4. What materials might be used to make a visible light LED?
5. List all the Group III and Group V elements.
6. Why are heterojunction designs used in optical sources?
7. What three properties of laser light make it different from ordinary light?
8. Which optical source would be used in a transatlantic communications system? Why?
9. What operational characteristics of optical sources would affect the dispersion loss in an optical fiber?
10. What properties of optical sources would affect the attenuation loss in an optical fiber?

Thought Questions

1. Obtain the manufacturer's data sheets for several LEDs or diode lasers. Compare the operating parameters of each. Describe what their advantages are.
2. Obtain the data sheets from two or more different manufacturers and compare how they describe and list the properties of their products.

3. Which type of light source would you think is most prevalent in fiber optic applications? Why?

4. What other technologies make use of LEDs and diode lasers?

5. Draw a schematic of a driver circuit for an LED and for a semiconductor laser using the pinouts from actual devices.

REFERENCES

Hecht, Jeff. *The Laser Guidebook,* 2nd ed., Blue Ridge Summit, PA: TAB books, 1992.

Senoir, John M. *Optical Fiber Communications—Principles and Practice,* Upper Saddle River, NJ: Prentice Hall International, 1985.

Sterling, Donald J. *Technician's Guide to Fiber Optics.* Albany, NY: Delmar Publishers, Inc., 1987.

Sze, S. M. *Semiconductor Devices—Physics and Technology.* New York: John Wiley & Sons, 1985.

Zanger, Henry, and Zanger, Cynthia. *Fiber Optics—Communications and Other Applications.* Upper Saddle River, NJ: Merrill/Prentice Hall, 1991.

6

OPTICAL DETECTORS

Included in this chapter:

6.1 INTRODUCTION

Whereas Chapter 5 considered those devices used to inject light into fiber, this chapter discusses those devices used to receive the light as it comes out of the fiber. The similarity in these devices is because both sources and detectors are based on semiconductor materials. Chapter 5 covered some of the concepts of semiconductors, and the references included at the end of this chapter also offer more insight.

Optical detectors, like their optical source counterparts, fall into one of two main designs. The **PIN photodiode** and the **avalanche photodiode** are really variations on the same theme. They do differ in operational characteristics and in basic design. The theory and operation of both devices are described in this chapter.

6.2 GENERAL THEORY OF OPERATION

Optical detectors are based on the p-n junction just as optical sources. The p-n junction of a detector is reverse biased (positive side of the power supply connected to the n-type material). Light that falls on the depletion region of the junction may supply enough energy to move an

81

electron from the valence band to the conduction band. This electron then passes through the circuit as current. Since the number of electrons liberated this way depends on the amount of light that falls on the depletion region, the current can be used as a measurement of the amount of light. The process of producing current using light is known as **photogeneration**. Mathematically, the current (I) is directly proportional to the amount of incident light (P) throughout most of the operating range of the detector, or

$$I \propto P \qquad (6.1)$$

Optical detectors (also known as photodiodes or photodetectors) operated in the reverse bias circuit just described are said to be **photoconductive**. This indicates that the conductivity of the detector is affected by the amount of light received. An increase in light incident (P) on the detector results in a decrease in the conductivity (ρ), or

$$\rho \propto I/P \qquad (6.2)$$

Detectors based on a p-n junction can also generate a voltage without the use of an external bias supply. In this **photovoltaic** operation, incident light liberates electrons which move toward the n-type material providing it with a negative charge. The absence of the electrons leaves holes behind so that there is a positive charge on the p-type material. The electrical potential between these charges is a voltage across the p-n junction. Current will flow if the diode is connected through a load resistance. The amount of voltage (and, subsequently, the amount of current) produced in this method is small relative to the values in the photoconductive mode. Because of this, photovoltaic operation is not normally used for communications systems.

6.3 OPERATIONAL CHARACTERISTICS

The major area of interest in operational characteristics of a photodetector is its response to incident light. Detectors that are sensitive to even low levels of light are useful because they allow a communications system to extend the distances covered by compensating for absorption loss. The response of a photodetector is described by its **responsivity**, **sensitivity**, and **quantum efficiency**.

The responsivity of a detector is the ratio of the current produced by the detector to the amount of light incident on its depletion region. The

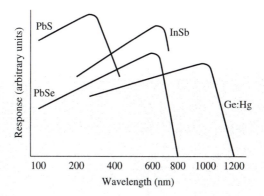

Figure 6.1 Wavelength Sensitivity for Common Detector Materials

ratio is normally expressed in μA (microamperes) to μW (microwatts) or amperes to watts. Mathematically, responsivity (R) is represented by

$$R = I/P \tag{6.3}$$

In a manner similar to semiconductor light sources, the response of a p-n junction detector is dependent on wavelength. As illustrated in Figure 6.1, the materials used to construct the detector will absorb only specific energy levels that are related to their band gap energy. This means detectors will respond to different wavelengths (or photon energies) of light in different ways. To express this wavelength sensitivity of a detector's response, manufacturers often supply the responsivity as a graph based on wavelength. As shown in Figure 6.2, the responsivity is a maximum (or peak) at a particular wavelength and tapers off above and below it. The term *sensitivity* is often used to describe the responsivity of a detector as a function of wavelength.

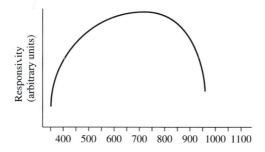

Figure 6.2 Responsivity as a Function of Wavelength

The quantum efficiency of a detector is an expression of its output in terms of particles. The ratio of the number of electrons moved to the valence band (N_e) to the number of photons incident on the detector (N_p) or

$$\eta = N_e/N_p \tag{6.4}$$

is the quantum efficiency (η).

A problem that results from using optical detectors is that some of the current in the circuit is not the result of the light but of other phenomena. This **noise**, as it is called, has several sources. Because of thermal energy in the components of a detector circuit (and especially the resistors) a certain number of electrons are released into the valence band without the aid of incoming light. These electrons contribute noise (or extraneous current) to the circuit. **Thermal noise** is also known as **Johnson noise** or **Nyquist noise**.

Thermal noise gives rise to current in the circuit even when no light is shining on the detector. Current that can occur without incoming light is known as the **dark current** of the detector.

Since the light is composed of photons, incident light on a detector can be visualized as a series of discrete particles similar to rain drops falling on a sidewalk. These photons arrive on the detector's surface at different times. Current, however, is normally measured as a steady stream (similar to a bucket of water being poured on a sidewalk), and this stream will fluctuate due to the particle nature of the incoming light. This fluctuation in current is known as **shot noise**.

To express the amount of noise in a detection circuit, the **signal-to-noise ratio (SNR)** is used. The SNR is a ratio of the current due to the incoming light (the signal or S) to the current due to the noise (N) or

$$\text{SNR} = S/N \tag{6.4}$$

A large signal-to-noise ratio means that the current due to the light is much greater than the current due to noise and is therefore easy to distinguish. A small signal-to-noise ratio means the noise current is close to the same amount as the signal current which is in danger of being masked or blocked out by the noise.

Several other quantities are used to express the amount of noise and the minimum amount of detectable light. Among these are the **minimum detectable power, detectivity, noise equivalent power (NEP)**, and **dee-star (D*)**. Each of these are essentially other methods of expressing the values already described. An adjustment in units used or the addition of wavelength or size factors leads to these pseudonyms.

Fortunately, most manufacturers provide information on how to use these characteristics to determine the suitability of a detector to a specific application.

The speed at which a detector can respond to changes in the amount of light incident upon it is also an important characteristic. As with light sources, detectors are characterized by their rise times and fall times or by their bandwidth. A factor in the speed of a detector is the amount of capacitance that it has in a circuit. The *RC time constant* (similar to the values used for capacitors in AC circuits) is sometimes used to express the speed of a detector.

The speed of a detector can also be considered in terms of its response to various frequencies of incoming light pulses. As a general trend, the output of the detector declines with an increase in frequency. The **cutoff frequency** is defined as the frequency at which the detector output drops to 3 dBs below its maximum output. Figure 6.3 illustrates the frequency response and cutoff frequency of a typical detector.

The *effective area* of a detector is important for determining the range of light (or numerical aperture) that a detector will pick up. Common detectors have a circular area which corresponds to the depletion region of the p-n junction. Their size is then expressed as the diameter of this circular region.

Detector packages are similar to those of sources. They are often equipped with microlenses or pigtails to facilitate connection to the end of a fiber. Some detectors are part of an integrated circuit which includes a preamplifier for boosting the signal from the detector. In addition to the standard diode/p-n junction, detectors can also be incorporated into transistors or darlington pairs.

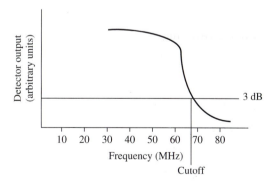

Figure 6.3 Cutoff Frequency

6.4 COMMON DETECTOR DESIGNS

One of the most common detector designs is the *PIN photodiode*. This detector incorporates an extra layer into the p-n junction (see Figure 6.4). This additional layer, known as the **intrinsic layer** (the *I* in PIN), is a neutral (neither p nor n) layer of semiconductor material that increases the size of the depletion region. An increased depletion region leads to an increased light-sensitive area allowing the detector to pick up a wider range of incoming light.

The intrinsic layer also serves to increase the speed of the PIN since the capacitance of the diode is reduced with an increased depletion region. Speed enhancement is also the result of reducing the number of photons that generate electrons in the p and n layers where they move more slowly.

PIN photodiodes typically provide 0.6 amps or more for every watt of incoming light. Their rise times vary between 5 and 50 ns, depending on the design and the bias voltage. The PIN is very inexpensive and can be operated with a fairly simple power supply and circuit. These properties make the PIN photodiode the most popular detector in optical fiber communications.

The *avalanche photodiode* (APD) is a variation on design that allows for increased responsivity. A strong electric field in the depletion region accelerates electrons in the conduction band. These accelerated electrons reach speeds high enough to cause additional electrons to break free into the conduction band when they collide with atoms (a process known as **photomultiplication**). The additional electrons are also accelerated and can cause the release of still more electrons. As a result, a large current can be produced with a small amount of incident light. Avalanche photodiodes often have responsivities on the order of 100 amps per incoming watt of optical power.

The rise time of an APD is comparable to most PIN diodes, but the drawback in the APD design is the need for a well-regulated bias voltage. The potential for photomultiplication means that noise current is also magnified, leading to a higher dark current and lower signal-to-noise ratio. In addition, a large bias voltage (a few hundred volts) is necessary to create the avalanche effect. These drawbacks make the APD

Figure 6.4 PIN Photodiode

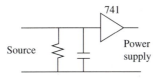

Figure 6.5 Simple Detector Circuit

more expensive to operate and limits its practicality to those applications (such as long-distance communications) where detection of low light levels is a necessity.

6.5 DETECTOR CIRCUITRY AND RECEIVERS

The signal produced by an optical detector must be amplified and processed before it can be used as a telephone/voice signal or a computer/digital signal. Amplification of the signal is normally accomplished electronically, although devices exist to amplify the optical signal before it is detected or as a method of extending the length of a communications line.

Electronic amplification incorporates fairly common operational amplifier circuits such as the low-impedance or transimpedance amplifier. Although it is possible to obtain a signal using a detector and a load resistor in a circuit such as the one shown in Figure 6.5, the operational amplifier method is preferred because of its faster operation.

The low-impedance amplifier circuit, shown in Figure 6.6, uses the load resistor as the negative input to a standard operational amplifier (such as a 741).

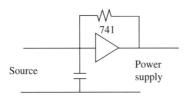

Figure 6.6 Low-impedance Amplifier

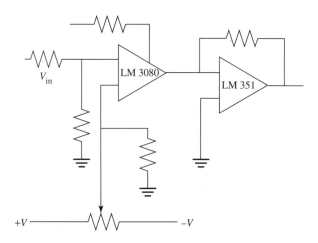

Figure 6.7 Transimpedance Amplifier

The voltage developed across the resistor is amplified by the circuit and can be adjusted by adjusting the ohmic value of the resistor. The low-impedance design suffers a drawback in operational speed since the RC time constant of the circuit increases with a greater load resistor design.

The transimpedance amplifier uses an open loop circuit which increases the gain of the amplifier by using a *feedback resistor* (see Figure 6.7). The response speed of this circuit is greatly increased but at the expense of gain which is a function of the feedback resistor size.

Signal processing in a detection circuit depends heavily on the type of communication used. Amplifier outputs can be standard TTL or other digital-based voltage levels which are then fed into logic circuits, or they may produce an analog signal for driving voice devices. In many applications, the fiber is not wired directly to the receiving device but is only one leg in the communications system. Detector circuits in these cases provide signals compatible with telephone switching and branching equipment or with distribution equipment for computer networks.

CHAPTER REVIEW

New Terms

Section 6.1
    PIN Photodiode
    Avalanche Photodiode

Section 6.2
> Photogeneration
> Photoconductive
> Photovoltaic

Section 6.3

Responsivity	Shot Noise
Sensitivity	Signal-to-Noise Ratio
Quantum Efficiency	Minimum Detectable Power
Noise	Detectivity
Thermal Noise	Noise Equivalent Power
Johnson Noise	Dee Star
Nyquist Noise	RC Time Constant
Dark Current	Effective Area

Section 6.4
> PIN
> Intrinsic
> APD
> Photomultiplication

Section 6.5
> Transimpedance Amplifier
> Low-impedance Amplifier
> Feedback Resistor

Review Questions

1. How are optical detectors similar to optical sources?
2. What affects the wavelength response of an optical detector?
3. What are the differences in characteristics between a PIN and an APD detector?
4. What is the function of the intrinsic layer in a PIN?
5. List the terms that might be used to describe the noise and sensitivity of a detector.
6. If you had to choose a detector for a computer network (short-distance communications), which detector would you choose? Why?
7. What affects the rise time of a detector? How can it be reduced?

Thought Questions

1. Explain the sources of noise in a detector. Which one is easiest to control? Which one has the least effect on the signal?

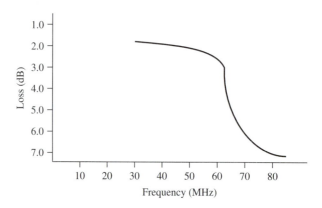

Figure 6.8 Diagram for Problem 4

2. Obtain manufacturer's data sheets for two detectors and compare their properties.

3. Using the information from Chapter 5, make a guess at what semi-conductor materials might be used to detect different wavelengths.

Problems

1. What would happen to the conductivity of a detector if the amount of incident light were doubled?

2. Calculate the effective area of a detector with a circular light-sensitive area with a diameter of 4 μm.

3. Calculate the quantum efficiency of an APD that produces 10 electrons for every one incident photon.

4. What is the cutoff frequency of the detector shown in Figure 6.8?

REFERENCES

Cherin, Allen H. *An Introduction to Optical Fibers.* New York: McGraw-Hill, 1983.

O'Shea, Donald. *Elements of Modern Optical Design.* New York: John Wiley & Sons, 1985.

Senoir, John M. *Optical Fiber Communications—Principles and Practice.* Upper Saddle River, NJ: Prentice Hall International, 1985.

Sterling, Donald J. *Technician's Guide to Fiber Optics.* Albany, NY: Delmar Publishers, 1987.

Sze, S. M. *Semiconductor Devices—Physics and Technology.* New York: John Wiley & Sons, 1985.

Zanger, Henry, and Zanger, Cynthia. *Fiber Optics—Communications and Other Applications.* Upper Saddle River, NJ: Merrill/Prentice Hall, 1991.

7

SPLICES, CONNECTORS, CABLES, AND OTHER COMPONENTS

Included in this chapter:

7.1 INTRODUCTION

To use optical fiber in communications, several devices are necessary for connecting the fiber to the source or detector and connecting one fiber to another. The fiber must also be packaged to withstand the environment in which it is installed. Optical fiber splices, connectors, and cables are a well-developed technology with several variations in cost, ease of use, and performance. The exchange, combination, and separation of signals in a fiber communications system requires the use of specialized devices as well. This chapter provides an informational discussion of these devices.

Installing fiber and associated devices requires specialized skills and knowledge much different from those required for installation of electrical cables. The discussions on installations in this chapter are meant to provide an overview of these skills and knowledge. Additional information is available from the various manufacturers of these products.

7.2 JOINING TWO FIBERS

Two fibers can be joined together with a **splice** or a **connector**. The splice is a permanent junction intended as a one-time-only joint which

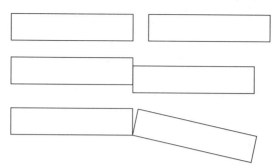

Figure 7.1 Extrinsic Loss

will not be undone. A connector acts as a temporary joint between fibers which can be undone and then rejoined many times. In both cases, misalignment of the two fibers and irregularity in the fiber construction can lead to a loss of light at the joint. This loss is defined in detail by the concepts of **extrinsic loss** and **intrinsic loss**.

Extrinsic loss is a result of a misalignment of the two fibers at the joint. The misalignment can be one of three types: **end separation, angular misalignment**, or **lateral misalignment**. Each type of misalignment (illustrated in Figure 7.1) increases the amount of loss exponentially as a function of the misalignment.

As shown in Figure 7.2, loss from the joint between the fibers is proportional to the log of the misalignment. The misalignment may be expressed as a distance (end separation and lateral misalignment) or as an angle (angular misalignment). The numerical aperture of the fibers involved is also a factor. Larger numerical apertures lead to lower losses because the light tends to cover a wider space and has a better chance of entering the adjoining fiber.

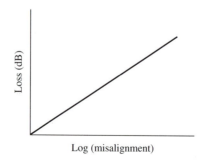

Log (misalignment)

Figure 7.2 Loss as a Function of Misalignment

Misalignment can also lead to loss due to reflection. Light exiting a fiber undergoes a Fresnel reflection at the glass/air interface as does light entering into a fiber. This reflection is approximately 5% at each place. Fresnel reflection loss can be compensated for through the use of an **index matching gel** that fills the gap between the fibers.

Intrinsic losses in a fiber joint are the result of irregularities or mismatches in the fiber rather than poor alignment by the connector or splice. The intrinsic losses are *numerical aperture mismatch, core diameter mismatch, cladding diameter mismatch, ellipticity,* and *concentricity*. The last three can be greatly reduced in the fiber manufacturing process whereas numerical aperture and core diameter mismatch are normally associated with connecting two different types of fibers.

A mismatch in numerical aperture can cause a loss when the fiber transmitting the light to the joint has a larger numerical aperture than the fiber receiving the light. In such a case, light spills out over the acceptance cone of the receiving fiber and is lost into the cladding or even into the surrounding air. Similar problems occur if the core diameter of the transmitting fiber is larger than the core diameter of the receiving fiber.

The cladding of a fiber is used as a guide to align two fibers in a joint. If there is a difference in cladding diameters, the alignment will be off. Generally, the cladding of one edge will be aligned, leaving a gap at the opposite edge as shown in Figure 7.3.

If the core of a fiber is not perfectly rounded, the light escaping it may not match the acceptance cone of the receiving fiber even though they both have the same numerical aperture. As illustrated in Figure 7.4, the long profile of an elliptical core can be perpendicular to the long profile of the other, leading to a loss of light. Additionally, in a connector, the alignment of these ellipses may change each time the two fibers are connected since their orientation can be changed.

If the core is not exactly in the middle of the fiber (or concentric with the cladding), there may be losses similar to the ellipticity problem. The cores of the two fibers would be shifted from each other, leav-

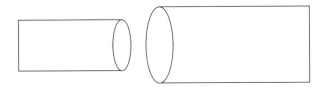

Figure 7.3 Cladding Diameter Mismatch

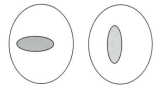

Figure 7.4 Ellipticity

ing a gap which could change if the fibers are disconnected and then re-connected with another orientation. Figure 7.4 illustrates this concen-tricity problem.

To reduce the amount of loss in a joint between two fibers, fiber manufacturers produce fiber within very tight specifications, and con-nector and splice designs align the fiber with microscopic accuracy. Fiber splice designs are categorized as either **mechanical** or **fusion** splices.

Mechanical splices use a housing that joins the fibers together with a minimum of misalignment and a mechanism that holds the fibers at a precise location within the housing. Two common examples are the **elastomeric splice** and the **four-rod splice**.

The elastomeric splice uses a rubber-like material (elastomer) with grooves cut into it. The grooves start fairly wide and then taper down to grip the fiber. The elastomer is housed within a glass sleeve that holds it in place and which is equipped with end guides that align the fiber with the grooves as it is inserted into the splice. The fiber is bonded into the splice with an adhesive that may also serve as an index matching gel. Figure 7.5 illustrates an elastomeric splice.

The four-rod splice uses four glass rods to hold the fiber in place. These rods curve outward slightly at the ends so that fiber can be in-

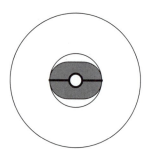

Figure 7.5 Elastomeric Splice

serted and then come together to form a gap into which the fiber fits. A steel sleeve holds the rods in place and is crimped over two silicon pieces that hold it to the fiber. Again, an adhesive/index matching gel holds the fibers in the splice.

Fusion splicing literally melts (or fuses) the two ends of a fiber together using a high-voltage electrical arc. A special piece of equipment, known as a *fusion splicer*, is used first to align the fibers and then to melt them together. Fusion splicers have become sophisticated forms of equipment and quite often have a video display which shows a magnified view of the two fibers. Splicers can also be equipped with a self-aligning system which injects light through the fibers and adjusts their position for minimum loss before fusing them. This feature also has the added benefit of displaying the loss due to the splice so that a technician can determine its quality.

Fiber connectors offer a trade-off between low loss and ease of use and installation. At the high-loss, easily used end is the **dry no polish (DNP)** connector which uses a simple catch to hold two fibers together and which is slid over the end of the fiber and held in place by plastic, v-shaped grooves.

For reduced loss, an *SMA connector* similar to the type used for connecting electrical cables is available. This connector has a cylindrical sleeve with a tapered nose and a tiny hole that holds the fiber. This piece, known as a *ferrule*, is epoxied to the fiber with a length of fiber extended past the hole. The ferrule is then placed into the SMA body so that the fiber sticks out past the end. The end of the fiber is then ground (or polished) until it is flush with the connector. The body of the SMA connector is threaded and screws onto a *bushing* which holds two fibers together. Figure 7.6 illustrates the components and assembly of an SMA connector.

The ends of a fiber must be prepared before they are inserted into a connector. To minimize loss, the ends must be polished so that they are

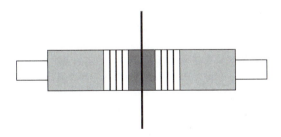

Figure 7.6 SMA Connector

smooth and perpendicular. Cracks or flaws in the surface of the fiber can scatter light, and sloped faces can cause misalignment. The preparation of the fiber ends has two steps: *cleaving* and *polishing*.

Before beginning the preparation of a fiber end, the outer protective coating found on most fibers must be removed. The coating, which is normally plastic, can be stripped from the fiber using a tool similar to wire strippers or removed using a solvent (such as acetone or paint thinner). The fiber is always cleaned after the coating is removed so that foreign particles do not obstruct the connection or splice.

In order to begin with a fairly flat and smooth surface, fibers are cleaved in a process similar to glass cutting. Because of the physical properties of glass, a scratch in the surface will cause it to break along the line of the scratch. To prepare an optical fiber, a hard, sharp **scribing tool** made of diamond or tungsten carbide is pulled across the fiber as it is pressed against its surface. (The scribing tool may be a simple blade or a more sophisticated tool that scribes and then stretches the fiber to break it.) The result is a scratch, and the fiber will break along this scratch when it is pulled from either side. A practiced technician can produce a clean, flawless fiber surface using this method, which is used to prepare fiber for splices as well as connectors.

For connectors, fiber ends are polished (sometimes without cleaving the fiber first) using at least two different grits of polishing paper or powder and generally with water applied for lubrication. Polishing can be done by hand or with a polishing machine. In either case, the fiber is moved through a figure-eight pattern to ensure even polishing. A prepared fiber end is always carefully inspected under a microscope before being used.

7.3 FIBER CABLE

To protect the fiber as it is routed underground, overhead, through walls, or under the ocean, several cable designs are available. Each design has different properties such as water resistance, flame retardant, crush proof, etc., but the components of the cables are roughly the same.

As shown in Figure 7.7, cable components, starting from the inside and moving outward, are **buffer**, **strength member**, and **jacket**. The buffer refers to both the plastic coating on the fiber itself and to an additional layer of plastic added when the fiber is manufactured into a cable. This second buffer can be a hard plastic tube with a diameter larger than that of the fiber allowing the fiber some movement. This is known as a *loose buffer*.

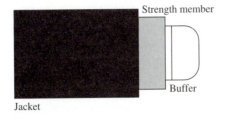

Strength member

Buffer

Jacket

Figure 7.7 Fiber Cable Components

A *tight buffer* is more of a coating around the fiber and is attached directly to the fiber's outer surface. Whereas loose buffer allows for expansion and contraction of the cable due to temperature fluctuations, the tight buffer offers more resistance to crushing or impacts.

The strength member of the cable absorbs the tensile and lateral stress that a cable may experience during installation or because of the bending, hanging, or compressing that may be required to fit it into its installation. Strength members may be a solid plastic or steel rod that runs through the middle of the fiber or may be strands (or yarn) of Kevlar® or a similar material. The type of strength member used will depend on where the cable will be located.

Jacket materials which encase the cable and protect all of its components are some type of plastic. Polyvinyl chloride (PVC) or polyethylene are used for cables intended primarily for indoor use in clean, dry environments such as ceilings or crawl spaces. Polypropylene, polyurethane, nylon, or Teflon™ offer good resistance to chemicals, heat, weather, and other conditions that may be found in a factory setting or in outdoor use.

Stress on the cable during installation and under normal operating conditions is an especially important consideration for cable design. Tensile stress (from pulling cable or from tension due to the cable's location and arrangement) and stress due to bending the fiber can lead to reduced performance and even permanent damage. Compressing the fiber cable by putting weight on top of it or by forcing it into a tight space can also lead to problems. Communications grade optical fiber is made of glass and can be brittle and delicate.

To avoid problems with a fiber cable, installation techniques and guidelines are provided by manufacturers. When installing fiber cable, minimum *bend radius* and *tensile force* must be considered. Fiber cable generally has a maximum tensile force allowed during installation; a lower maximum tensile force is allowed once the cable is installed. To monitor this force, a mechanical gauge (also known as a *tensiometer* or

dynamometer) is used while the cable is pulled into place and to check it after it is in place. The cable is always pulled by its strength member so that a minimum amount of stress is placed on the fiber itself.

The minimum bend radius of the cable is the limit beyond which the fiber may break or crack. Cable routes must be carefully evaluated to ensure that this radius is not missed, causing fiber damage during installation or use. Cable is sometimes installed in conduit to limit the bends it makes and to protect it from damage due to crushing forces or chemical contaminants.

Various types of cable designs are available for the many environments in which it is used. *Undercarpet* cable is placed in a flattened-out pvc jacket and includes plastic guides for bends to hold the fiber in place and keep it within the minimum bend radius. Cables used behind walls, in ceilings, and under false floors (areas known as **plenums**) meets the National Electrical Code requirements for fireproof or smoke-free materials.

Outdoor cables, such as those on telephone poles or buried in trenches, are often equipped with a protective steel sheath to protect against rodents that sometimes chew on the cable. Sometimes, a gel is used within the cable to expunge any air and limit damage due to freezing or moisture. Some cables include a copper wire that allows installers to send communications along the cable during installation and testing. Fiber/copper cables are known as **hybrids**.

Fiber cable may contain a single fiber or multiple fibers. Multiple fiber cables with two fibers are known as **duplex** cable, and single fiber cables are called **simplex**. These terms refer to two common communications methods (see Chapter 9 for more details). Multiple fibers may be arranged side by side, in a circular position around the strength member, or wrapped around the strength member in a helical pattern. The strength member often has grooves for holding the fibers in their correct positions.

7.4 COUPLERS AND SWITCHES

A **coupler** is used to combine or separate signals from multiple fibers. In its most basic form, a coupler is a device that connects two or more fibers together without terminating the fibers. As illustrated in Figure 7.8, the coupler can be thought of as an intersection of two roads. Cars traveling into the intersection may continue on through or turn down one of the other roads. Two cars may come from different directions and then travel the same road after the intersection.

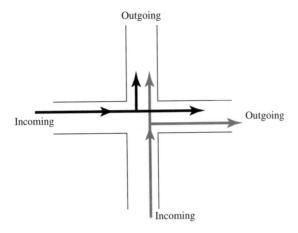

Figure 7.8 Coupler Principles

To accomplish fiber coupling, the simplest method is to fuse two fibers into a *T coupler*. This coupler allows two fiber signals to be combined onto one fiber or a single fiber signal to be split into two parts. The split signal may be equal or may be some ratio of optical power (for example, 60% of the signal stays in the original fiber, and the other 40% is split into the new fiber). This second case is often used in a network where a portion of the main signal must be sent to several locations.

T coupling may also be accomplished by a set of optical components as shown in Figure 7.9. In this case (known as an *expanded beam cou-*

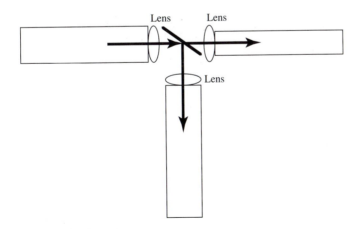

Figure 7.9 Expanded Beam T Coupler

pler), the light from the fiber is expanded through the use of a lens. A beam splitter splits the light into two parts, and the light is then focused by lenses into two fibers.

The use of a **GRIN** (*graded index*) lens may also accomplish T coupling. In this case, two fibers are fed into one end of the lens which focuses light into a signal fiber on the other end. GRIN lenses differ from ordinary lenses because they focus light through changes in the index of refraction rather than through curving the surface of the lens.

When light into a coupler should be divided equally into several fibers, a **star coupler** is used. The star coupler may use a **mixing block** which disperses light from one fiber into all the other fibers in a coupler. Another design uses a **reflective surface** which redirects light by reflection into the fibers. Star couplers offer the advantage of making a split into several fibers in one coupler rather than using a combination of T couplers, each of which would cause loss in the system.

The loss caused by a coupler is an important consideration in coupler design. The loss due to a coupler is divided into four categories: *throughput loss, tap loss, directional loss*, and *excess loss*. Throughput is a term used to describe the light that continues along the original fiber after the coupler. If we consider a coupler with two fibers (the fiber that delivers the signal and the fiber which is receiving the signal), then the coupler will actually have four connections (or *ports*). Two represent the incoming signal fiber and the incoming receiving fiber, and two represent the outgoing signal fiber and the outgoing receiving fiber (see Figure 7.10).

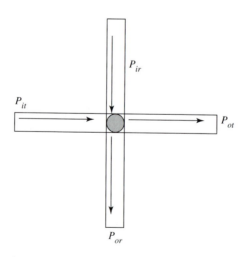

Figure 7.10 Coupler Loss

The throughput loss of a coupler is the amount of light removed from the incoming signal. Ideally, all of the light removed is channeled into the other fiber in the coupler, but, in reality, some of it is lost.

Some of the light lost in the coupler appears on the incoming portion of the receiving fiber and travels backward down that fiber. This light is accounted for by the directional loss. The rest of the light is channeled into the outgoing receiving fiber and is accounted for by the tap loss.

Some light is also lost through scattering, reflection, and other mechanisms in the coupler. This lost light does not appear at any of the fiber ports. The light lost in this way, along with the light lost in the directional loss just described, is accounted for by the excess loss of the coupler.

If we use the designations P_{it} for the incoming transmitting fiber power, P_{ir} for the incoming receiving fiber power, P_{ot} for the outgoing transmitting fiber power, and P_{or} for the outgoing receiving fiber power, the four loss types can be calculated by the equations

$$\text{(throughput loss) } L_t = 10 \log(P_{ot}/P_{it}) \tag{7.1}$$
$$\text{(tap loss) } L_p = 10 \log(P_{or}/P_{it}) \tag{7.2}$$
$$\text{(directional loss) } L_d = 10 \log(P_{ir}/P_{it}) \tag{7.3}$$
$$\text{(excess loss) } L_e = 10 \log((P_{ot} + P_{or})/P_{it}) \tag{7.4}$$

There is also some loss caused by the connection of the fibers to the coupler itself. Example 7.1 illustrates how to calculate the loss of a fiber coupler.

 EXAMPLE 7.1

A fiber coupler has the following power levels: Incoming transmitting fiber is 10 mW; outgoing transmitting fiber, 8 mW; incoming receiving fiber, 0.5 mW; outgoing receiving fiber, 1 mW. Calculate the throughput loss, tap loss, directional loss, and excess loss of the coupler.

To calculate losses

Using Equations 7.1 through 7.4,

$$\begin{aligned}
\text{Throughput Loss} \quad L_t &= 10 \log(P_{ot}/P_{it}) \\
&= 10 \log(8/10) \\
&= -0.97 \text{ dB}
\end{aligned}$$

$$\text{Tap Loss} \qquad L_p = 10 \log(P_{or}/P_{it})$$
$$= 10 \log(1/10)$$
$$= -10 \text{ dB}$$

$$\text{Directional Loss} \qquad L_d = 10 \log(P_{ir}/P_{it})$$
$$= 10 \log(0.5/10)$$
$$= -13 \text{ dB}$$

$$\text{Excess Loss} \qquad L_e = 10 \log((P_{ot} + P_{or})/P_{it})$$
$$= 10 \log((8 + 1)/10)$$
$$= 10 \log(9/10)$$
$$= -0.046 \text{ dB}$$

A special type of coupler used with optical fiber, known as a **wavelength division multiplexer** (**WDM**), allows multiple signals to be transmitted down a fiber simultaneously. Each signal uses a different wavelength of light and is, therefore, distinguishable from the other signals in a fiber. To accomplish this, a WDM separates the signals by wavelength using diffraction or interference.

In a *diffraction WDM*, the coupler uses a diffraction grating (see Chapter 2) which separates the signals by reflection. Incoming light reflects from the grating at different angles, depending on wavelength. The outgoing fibers can then be positioned to intercept the separate wavelengths.

Interference coatings in a WDM coupler cause reflection of light at one wavelength and transmission of light at other wavelengths. The reflected light is constructively interfered (see Chapter 2) due to the thickness of a layer of **dichroic material** deposited on the end of the fiber. A fiber can then be positioned to receive this reflected light and separate it from the rest of the signal. If additional signals must be separated, additional coatings must be used.

Because the WDM devices in use today are only efficient when the separate signals have a wide difference in wavelength, only two or three signals are normally multiplexed. These signals are at common fiber optic wavelengths such as 820 nm, 1300 nm, and 1550 nm.

Switches used in fiber optic communications allow light in a fiber to be redirected into another fiber or to continue in the same fiber depending on the setting. As with electrical switches, optical fiber switches choose between two possible paths for the signal. Switches may by *mechanical* or *optical* in nature.

Figure 7.11 Mechanical Switches

Mechanical switches may move a transmitting fiber between one of two possible receiving fibers through the use of an electromechanical device such as a solenoid. Another option uses a prism that is rotated so that light emerging from it is directed at one of two fibers. These two types of switches are illustrated in Figure 7.11.

Because mechanical switches are relatively slow devices, optical switches have been invented to speed up the process. An optical switch works because some materials change their index of refraction when a voltage is applied to them. These *electro-optical* materials are constructed into an optical element that refracts light toward one fiber when no voltage is applied and to another fiber when the voltage is turned on. Such switching does not require any physical movement and is inherently faster than mechanical methods.

CHAPTER REVIEW

New Terms

Section 7.2

Splice

Connector

Extrinsic Loss

Intrinsic Loss

End Separation

Angular Misalignment

Lateral Misalignment

Index Matching Gel

Numerical Aperture Mismatch

Core Diameter Mismatch

Cladding Diameter Mismatch

Ellipticity

Concentricity

Mechanical Splice

Fusion Splice

Elastomeric Splice

Four-rod Splice

Fusion Splicer

Dry No Polish (DNP)

SMA Connector

Ferrule

Busing

Cleaving

Polishing

Scribing Tool

Section 7.3

Buffer	Tension
Strength Member	Dynamometer
Jacket	Undercarpet Cable
Loose Buffer	Plenum
Tight Buffer	Hybrid
Bend Radius	Duplex
Tensile Force	Simplex

Section 7.4

Coupler	Excess Loss
T Coupler	Wavelength Division
Expanded Beam T Coupler	Multiplexer (WDM)
Graded Index (GRIN) Lens Coupler	Diffraction WDM
Star Coupler	Interference WDM
Mixing Block Star Coupler	Dichroic Material
Reflective Surface Star coupler	Switches
Throughput Loss	Mechanical Switches
Tap Loss	Optical Switches
Directional Loss	Electro-optic Materials

Other Chapters with Related Information

Chapter 8
Chapter 9

Review Questions

1. What are the sources of loss in a splice or connection? How are they reduced?
2. What is the difference between a splice and a connector?
3. Describe the elastomeric splice and the four-rod splice.
4. A fiber is used to connect the computers in a network that is all in one room. What type of cable might be used?
5. What is a plenum?
6. If a cable contains more than one fiber, how might they be arranged?
7. What types of materials are used in cable jackets?
8. How must a fiber end be prepared for connection to another fiber?
9. What is the simplest and cheapest type of connector?
10. What two things must be taken into consideration when installing fiber cable?

Thought Questions

1. Explain how and why the following would affect the loss caused by a connector.
   a. The orientation of the two halves of the connector.
   b. The cleanliness of the ends of the fiber.
   c. Burrs or cracks on the surface of the fiber.
   d. Type of epoxy used.

2. Discuss the trade-off between using a connector for a joint that may be removed only under special circumstances (say, for troubleshooting or repair) and using a splice.

3. Name some applications that are likely to use several fiber switches. Explain why.

REFERENCES

Cherin, Allen H. *An Introduction to Optical Fibers.* New York: McGraw-Hill, 1983.

O'Shea, Donald. *Elements of Modern Optical Design.* New York: John Wiley & Sons, 1985.

Senoir, John M. *Optical Fiber Communications—Principles and Practice.* Upper Saddle River, NJ: Prentice Hall International, 1985.

Sterling, Donald J. *Technician's Guide to Fiber Optics.* Albany, NY: Delmar Publishers, Inc., 1987.

Sze, S. M. *Semiconductor Devices—Physics and Technology.* New York: John Wiley & Sons, 1985.

Zanger, Henry, and Zanger, Cynthia. *Fiber Optics—Communications and Other Applications.* Upper Saddle River, NJ: Merrill/Prentice Hall, 1991.

8

FUNDAMENTALS OF COMMUNICATIONS

Included in this chapter:

8.1 INTRODUCTION

Since the emphasis of this text is on optical fiber used in communications, this chapter defines some of the principles of communications, especially as they relate to optical fiber. This overview assumes some basic knowledge of AC electronics and digital electronics. Further information about these subjects can be found in the references listed at the end of the chapter.

Communications can be fundamentally divided into two areas: **analog** and **digital**. Analog refers to a continuous, smooth change of light in a fiber. One example of an analog device is a standard clock with a numbered face and minute and hour hands. The hands of the clock change in a smooth flow rather than in steps. In other words, the minute hand rotates around the clock face continuously and does not jump from number to number.

A digital system consists of discrete, distinguishable steps. Consider the digital clock on your VCR, clock radio, or digital watch. The time display changes in abrupt steps (from 12:00 to 12:01 to 12:02, for example), and there are definable gaps between each display. Digital is sometimes further defined in terms of two possible steps or conditions. These two possible states are known as 1 and 0, on and off, or true and false and can be used to represent any quantity when several of them are combined.

109

Digital representation of values (or **encoding**) uses a series of 1's and 0's to represent certain numbers or even letters. A good analogy is the Morse code system which represents each letter in the alphabet as a series of dots and dashes. Each combination of dots and dashes can be interpreted as an A, B, C, etc. By the same token, each combination of 1s and 0s in a digital system can be interpreted as a number, letter, or a special code in a communications system.

In an historical context, communications began as a strictly analog process. Original radio, telegraph, and even telephone systems were analog simply because digital did not exist. Analog communications had an advantage since all information transmitted at that time (voice, music, and other sounds) was analog in nature.

With the advent of transistors in the middle of this century, the possibility of digital communications was born. As solid-state electronics evolved into integrated circuits and other devices, they showed an inherent advantage in speed, size, and power consumption when compared to traditional analog devices. Digital electronics led to the development of desktop and laptop computers, handheld calculators, and miniature radios. Digital became king because it could "out-do" analog in so many ways.

Still, the majority of information handled by these devices was analog in nature. Telephones (for awhile, anyway) still were used primarily to transmit voices. Radios and TVs dealt with sound and video (analog quantities). As a result, communication systems such as telephone lines, TV and radio signals, and even cable TV still maintained their analog nature.

Digital communications began to gain ground as more information being transferred became digital in nature. Computer-to-computer communications is the primary example. Because of the advantage of using digital equipment (speed, size, and power consumption), certain aspects of communications systems were converted to digital.

Today, telephone systems (and associated systems such as computer communications, through telephone lines and cable TV) are a hybrid of digital and analog. The technology at the ends of the system (your telephone for example) is still analog, at least in terms of the signal sent out (although your telephone is probably built of digital components, it sends an analog signal to the phone lines). However, your local telephone switching station and the long-distance lines, which are in between the ends, are purely digital.

This hybrid approach is rapidly diminishing as the telephone company and others replace analog with digital. Soon the system will be completely digital from end to end. Therefore, this chapter briefly discusses analog communications but spends more time on digital. Some information about the conversion between the two is also included.

8.2 ANALOG COMMUNICATIONS

To communicate in an analog system, the properties of a wave-shaped carrier are modified to represent information that needs to be transmitted. The two simplest methods for accomplishing this are **amplitude modulation** (**AM**) and **frequency modulation** (**FM**). Amplitude modulation changes the amplitude of a carrier wave to represent communication data. The carrier amplitude increases and decreases at a rate that corresponds to the frequency of the analog signal of the communications data. In such a system, as with all analog communications, the wave that carries the information is known as the **carrier signal** or *frequency*. The data that is communicated is known as the **intelligence signal** or *data signal*, or *frequency*.

Frequency modulation changes the frequency of the carrier signal at a rate that corresponds to the changes in the frequency of the intelligence signal. Whereas AM can be considered the compression of the carrier wave in the vertical direction, FM is the compression in the horizontal. Figure 8.1 illustrates the difference between AM and FM.

To accomplish analog communications in optical fiber, the carrier (light) must be modified. Amplitude or frequency modulation of the light wave itself is impossible because of the extremely high frequencies of light waves. Instead, a lower frequency signal is imposed on the light wave to create the carrier wave. This signal is then modulated to send data.

Imposing and modulating the carrier frequency on the light in a fiber is accomplished via the power supply for the optical source. By adjusting the voltage up or down, a transmitting circuit can adjust the optical power up and down. The result is an oscillating output from the optical source. These oscillations can then be modified by adjusting the

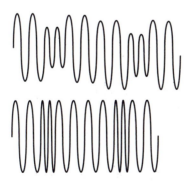

Figure 8.1 AM (top) and FM (bottom) Signals

amount of voltage (AM) or the frequency at which the voltage changes are made (FM).

8.3 DIGITAL COMMUNICATIONS

Digital communications involve producing a series of pulses in the light output of the source. This method (sometimes known as **pulse code modulation** or (**PCM**) is used to represent the 1s and 0s of a digital signal. In simplest terms, the 1s and 0s could be represented by a presence or absence of pulses. For example, a pulse in the light output would indicate a 1 and a lack of pulse indicates a 0. Unfortunately, two problems exist with this basic method. First, sending a chain of 0s, one right after the other, could lead to confusion at the other end. Does the prolonged absence of pulses indicate one 0 or several, and if several, how many? Even more discouraging, a complete absence of pulses may mean no signal is being sent at all. To solve this problem, a timing system must be added to the communication.

Communication timing can be **synchronous** or **asynchronous**. Synchronous communications means the receiving end and the transmitting end both operate under the same timing. The timing is supplied by a clock signal that is transmitted with the communications data. The transmitting end times each pulse (or lack of pulse) with this signal so that there is a set time for each one. At the receiving end, the detection system splits the incoming signal into pieces of the same size using the accompanying clock signal.

In asynchronous communications, no clock signal is used to link the timing of both ends of the system. Instead, the transmitting system sends a prearranged series of pulses to warn the receiving end that a signal is about to be sent. Once the transmission begins, the pulses (or lack of pulses) are still timed as before, but each end has its own clock signal to use in dividing up the signal.

The second problem with the basic method of sending a pulse for a 1 and a lack of pulse for a 0 comes from possible distortions to the signal as it travels through the system. As discussed in Chapter 4, fiber signals are distorted by many mechanisms, any one of which could lead to a change in pulse length, overlapping of pulses, or even a loss of pulses altogether. These possible distortions would make the receiving end misinterpret the signal. Two solutions to this problem exist: **error detection** and *digital encoding systems*.

One way of overcoming the possible distortion problem is to devise a system by which the receiving end has at least a partial idea of what the signal should be. That way, any errors in signal will be apparent, and the receiver can act accordingly. Generally, the receiver will either indicate that there is an error so that any type of equipment to which it is connected will be able to compensate for the error, or will ask for a complete retransmission of the signal.

Checking for errors is often accomplished using the **parity** system. Parity refers to the number of 1s in a standard digital transmission. Data is normally sent in groups (or **words**) of 1s and 0s. A single 1 or 0 is known as a **bit**, four would be a **nibble** and eight would be a **byte**. Some communication systems may use 16, 32, or more bits in a word. Whatever the case, a word with an even number of 1s is said to have *even parity*. An odd number of 1s is known as *odd parity*.

In a parity error checking system, each word has an additional bit added to make it have a set parity. For example, an odd parity system would add an extra bit (1 or 0) so that the total number of bits is odd. At the receiving end, the parity of each word can be checked, and if a word has the wrong parity, an error in the transmission occurs. Figure 8.2 illustrates how even and odd parity would be achieved.

Parity error checking offers a simple way to check for errors in transmission; however, it has its faults. Since it is possible that two bits could get changed in a transmission, the parity of a word that contains an error may still be correct. To solve this problem, an even more advanced system is used. This system is known as **cyclic redundancy check (CRC)**.

The CRC uses a mathematical formula to detect errors. Each data word (which is a number, in a sense) is passed through this formula, and the result is transmitted along with the word. At the receiving end, the same process is repeated, and the result is compared with the transmitted result. Since the odds of the words being distorted so that they produce the same result are slim, CRC is a very effective method for checking for errors.

			Final Word	
Data	Even Parity Bit	Odd Parity Bit	Even	Odd
1101	1	0	11011	11010
1111	0	1	11110	11111
1001	0	1	10010	10011
0001	1	0	00011	00010

Figure 8.2 Parity

The second solution for detecting errors in transmission is the use of advanced digital encoding systems. These systems deviate from the pulse = 1 and no pulse = 0 method of encoding so that the possibility for error is reduced. There are many such encoding methods, but the most popular are *nonreturn to zero, return to zero, nonreturn to zero inverted, manchester, miller,* and *biphase*.

The **return to zero** (RZ) code uses the absence of light to indicate a 0. If a 1 is transmitted, the light is turned on for half the length of a pulse and then returns to 0 (no light) for the rest of the length. For each 1 sent, the signal always returns to 0 before the next bit.

The **nonreturn to zero** (**NRZ**) code uses the presence of light to represent a 1 and the absence of light to represent a 0. The light then pulses off and on for 1s and 0s, but a chain of 1s (or 0s) does not produce a change in the light output. In other words, the light would stay on continuously (not return to 0) for a series of three 1s.

In the **nonreturn to zero inverted code** (**NRZI**), a change in the light output (either the light turning on or the light turning off) represents a 0. No change in the light output represents a 1. In this code, a string of three 0s would be represented by three changes in the signal (for example on, off, on), and a string of three 1s would be represented by no changes in the signal. Return to zero, nonreturn to zero, and nonreturn to zero inverted are illustrated in Figure 8.3.

The **Manchester code** has a change in the light output for each pulse period. For a 1, the first half of the period is high, and the second half is low. When the first half of the period is low and the second half is high, the bit is a 0. The **Miller code** is similar except that a 0 is represented by no change if it follows a 1 or by a change at the beginning of the period if it follows another 0. In the **biphase code**, each period

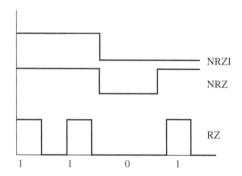

Figure 8.3 Return-to-Zero, Nonreturn to Zero and Nonreturn to Zero Inverted Codes

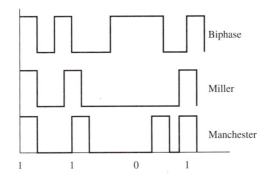

Figure 8.4 Manchester, Biphase, and Miller Codes

begins with a change. For a 1, the signal changes again midway through the period, and for a 0, no change occurs. All three codes are illustrated in Figure 8.4.

8.4 ANALOG-TO-DIGITAL AND DIGITAL-TO-ANALOG CONVERSIONS

Because of the hybrid nature of modern communications and because some data begins as analog and some as digital, a method is needed to convert analog signals to digital (*a to d* or **A/D** conversion) and to convert digital signals to analog (*d to a* or **D/A** conversion). The methods used are universal to other applications such as recording and voice generation as well as to communications.

To convert an analog signal to a digital signal, the signal must be divided into sections known as **samples**. Each sample is then represented by a digital code. Smaller samples and more bits in the code increase the accuracy of the conversion. As a minimum, the sampling rate must be twice the frequency of the original analog signal. A signal of 5000 Hz, for example, would have to be sampled 10,000 times per second to produce an accurate digital version of the signal. To convert the digital signal back to an analog signal, each digital code is changed to the signal level it represents, and the entire signal is reconstructed.

Several methods are used to accomplish the sampling/conversion process. In the **simultaneous** or **flash** method, the entire analog signal is sampled and converted at one time. This process becomes cumbersome for even moderately large signals, and a more sequential method is commonly used instead. For example, the *stairstep-ramp* method takes each sequential part of the signal and steps through the possible digital codes until it finds one that matches the analog signal level.

To increase conversion speed, the *tracking* method begins in the same way as the stairstep-ramp, but instead of resetting to zero and counting up to the necessary value again, the system uses the last value as a starting point. The digital code is then increased or decreased from that point until a match is found. Since the difference between successive samples is normally small, this method encodes the signal much quicker.

The *slope* method follows a set increase in voltage (or slope) over a period of time. The converter counts through digital codes as the voltage increases, and when a match is found, the final count is used as the encoded digital signal.

Several possible systems can be used to reconvert the digital signal to analog. Of these, the *binary weighted input* and the *R/2R ladder* are common. The variations between these methods are mostly electronic in nature, but they are both based on using various resistor values to produce the proper voltage for an incoming digital code.

8.5 MULTIPLEXING AND DEMULTIPLEXING

In many cases, the immense capacity of an optical fiber is underused by standard communications, especially those that are primarily voice or telephone. To make use of this excess room, a system of **multiplexing** and **demultiplexing** is used. Put simply, multiplexing allows several signals to be sent down a fiber simultaneously. Demultiplexing involves dividing these signals into their individual parts at the receiving end of the communications system. Multiplexing/demultiplexing can be accomplished by *wavelength division* (discussed in Chapter 7) or by *time division*.

Wavelength division multiplexing uses a separate wavelength of light to carry each signal down the fiber. The various wavelengths are normally far enough apart that they can be distinguished easily at the receiving end. Because of this requirement, the number of wavelengths available to carry signals is limited to two or three. This means wavelength division multiplexing can carry only a small number of "extra" signals.

Time division multiplexing makes use of a "select and send" system like the one illustrated in Figure 8.5. Several signals are sent down the fiber serially, one piece at a time. A portion of signal *A* in the diagram is sent down the fiber followed by a portion of signal *B* and a portion of *C*.

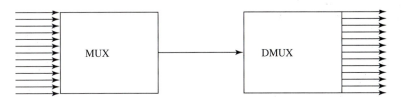

Figure 8.5 Time Division Multiplexing/Demultiplexing

The next portion of signal *A* is then extracted and sent followed by *B* and then *C* as the pattern repeats itself. At the receiving end, the signals are reconstructed, piece by piece. The process of time division multiplexing is extremely fast so that the transmitter and the receiver do not perceive any gaps or waiting periods in the transmission.

CHAPTER REVIEW

New Terms

Section 8.1
Analog
Digital
Encoding

Section 8.2
Amplitude Modulation (AM)
Frequency Modulation (FM)
Carrier Signal or Frequency
Intelligence or Data Signal or Frequency

Section 8.3

Pulse Coded Modulation	Even Parity
Synchronous	Odd Parity
Asynchronous	Cyclic Redundancy Check (CRC)
Error Detection	Nonreturn to Zero (NRZ)
Digital Encoding Systems	Return to Zero (RZ)
Parity	Nonreturn to Zero Inverted (NRZI)
Word	Manchester
Bit	Miller
Nibble	Biphase
Byte	

Section 8.4

Analog to Digital (A/D)	Stairstep-Ramp
Digital to Analog (D/A)	Tracking
Sample	Slope
Simultaneous	Binary Weighted Input
Flash	R/2R Ladder

Section 8.5

Multiplexing
Demultiplexing
Wavelength Division
Time Division

Chapters with Related Information

Chapter 1
Chapter 7

Review Questions

1. Name two analog and two digital quantities found in everyday life.

2. Why are modern communications systems a hybrid of analog and digital?

3. In what way is a carrier wave modified in AM communications? In what is it modified for FM communications?

4. What are the two states of a digital system?

5. How can errors in transmission be avoided and detected in a digital communications system?

6. What would be the parity of the following data words?
   a. 11100111
   b. 1001
   c. 01

7. What sampling rate would be required for a 1500 Hz analog signal that is to be converted to digital?

8. Name two codes that would have a change in the light power every time a 1 is transmitted.

9. Do you think the telephone company multiplexes long-distance telephone calls? Why or why not?

10. What method is used to modulate an optical signal to represent an analog signal?

Thought Questions

1. Which type of optical source (laser or LED) would be best suited for analog communication? Why?
2. Describe a typical analog-to-digital conversion application.
3. Explain how multiplexing would help in other applications besides telephone communications.
4. Using a table of binary codes (such as the ASCII system), encode the following message in binary and then add parity bits.

Testing 1 2 3.

REFERENCES

Cherin, Allen H. *An Introduction to Optical Fibers.* New York: McGraw-Hill, 1983.

O'Shea, Donald. *Elements of Modern Optical Design.* New York: John Wiley & Sons, 1985.

Senoir, John M. *Optical Fiber Communications—Principles and Practice.* Upper Saddle River, NJ: Prentice Hall International, 1985.

Shrader, Robert. *Electronic Communication.* New York: McGraw-Hill Book Company, 1990.

Sterling, Donald J. *Technician's Guide to Fiber Optics.* Albany, NY: Delmar Publishers Inc., 1987.

Zanger, Henry, and Zanger, Cynthia. *Fiber Optics—Communications and Other Applications.* Upper Saddle River, NJ: Merrill/Prentice Hall, 1991.

9

COMMUNICATIONS APPLICATIONS FOR OPTICAL FIBER

Included in this chapter:

9.1 INTRODUCTION

Optical fiber is an ideal communications medium for systems that require high data capacity, fast operating speed, and/or long distance with a minimum number of repeaters. These properties have been historically required by long-distance telephone systems and computer networks. Over time, however, the other advantages of fiber (such as size and weight, imperviousness to electromagnetic interference, etc.) and the falling cost of materials and installation have led to its use in medium-length telephone lines and the first stages of connection between residences and businesses and the phone company (known as "fiber to the home"). This last case means an infinite number of possibilities to the average consumer, including **high-definition television** (**HDTV**), expanded cable television with the possibility of a channel for every viewer, and the highly vaunted Information Superhighway.

This chapter describes the two main applications—long-haul telephone and computer networks—which provides a fairly complete picture of fiber as a communications tool. Some other areas are also covered to give more general information on the use of fiber in other applications besides communications. Fiber detectors, medical tools, and video systems are widely used and offer many advantages. Information on these applications can be found in the references listed at the end of this chapter.

121

Understanding fiber communications applications first requires interpreting the myriad of acronyms, catch phrases, and other jargon associated with it. Table 9.1 gives a brief definition of some of the more common terms.

Table 9.1 Fiber Jargon

Phrase	Definition
POTS	Plain Old Telephone Service—Standard telephone (voice) communications
FDDI	Fiber Data Distributive Interface—A standard established to help conform fiber communication systems for computers.
Category 5	A standard for fiber/telecommunications equipment performance parameters. Used to help rank this equipment.
CATV	Cable television
HDTV	High-definition television—A new television system that promises high picture and sound quality.
ISDN	Integrated Services Digital Network—Standards used to define transmission of voice, data, and video which are services that fiber can deliver to residences and businesses.
POF	Plastic optical fiber
PCS	Plastic clad silica—Fiber with a glass core and a plastic cladding.

9.2 TELEPHONE COMMUNICATIONS

To better understand the use of optical fiber in telephone communications, a basic overview of the telephone system in general is required. As illustrated in Figure 9.1, telephone companies have three levels of communications. The phone line from your home or business is connected to a common point that serves a neighborhood or similar geographic area and is called the *servicing area interface*. Several servicing

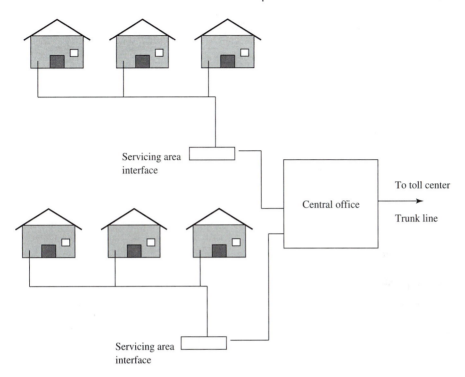

Figure 9.1 The Telephone System

area interfaces then feed into a *central office* which is designated by a *prefix* or *exchange*, the first three numbers of a telephone number (not the area code). Central offices are then connected to *long-distance toll centers* which are connected to other area codes. The toll center is a clearinghouse for all long-distance calls traveling in or out of a *local access transport area* (LATA).

Originally, optical fiber was used in the **trunk lines** that feed long-distance calls from the toll centers into a central office. As the cost of fiber has dropped, it has been used in the connections between a central office and the local service area interfaces. Some communities in a few experimental areas have fiber directly to houses.

Telephone signals are characterized as **voice channels** which are essentially two-way telephone conversations. Basic telephone signals (known as *DS1* or *T1* signals) contain 24 voice channels which can pass through a copper wire system efficiently. Several DS1 signals are then multiplexed into a fiber line for transmission to other locations. In the late 1970s, another standard telephone signal known as *DS3* (or *T3*) was defined, having the capacity for 672 voice channels and intended for use with fiber lines. Since that time, two other signals have been defined. These signals, known as *T3C* and *T4*, can have 1244 and 4032 voice channels, respectively.

The signals transmitted in telephone lines is normally pulse code modulated and can be rated in terms of the number of kilobits per second (kb/s or kbps) transmitted. Data rates are all multiples of 64 kb/s, starting with D1 at 1.544 megabytes per second (Mb/s or Mbps) and extending through T4 at 274.176 Mb/s. Other signals at higher data rates do exist and have been used in some applications.

A typical fiber optic system would be similar to the one shown in Figure 9.2. This system accepts the copper wire feed from several residential lines at a servicing area interface. Each line (24 voice channel) is fed into one of two multiplexers each capable of accepting up to 28 incoming DS1 signals. The signals are then multiplexed into a T3C (roughly 1300 voice channels at 90 Mb/s) which routes the signals to the central office. The signals are also expanded to include error-checking information which is then removed at the receiving end. At the central office, these signals are redistributed into other multiplexers (at a higher capacity) and sent to their appropriate destinations.

An example of a telephony application of fiber optic communication is a T3C transmission system. This system operates at 90 Mb/s and handles 1344 voice channels. As illustrated in Figure 9.2, the T3 system would be typically fed by multiplexed T1 lines. Since T1 lines provide 24 voice channels, the T3C would recive 56 (1344 ÷ 24) T1 lines. These lines would normally be fed through two multiplexers handling 28 lines each. Some of the lines and the bits in the incoming signals are re-

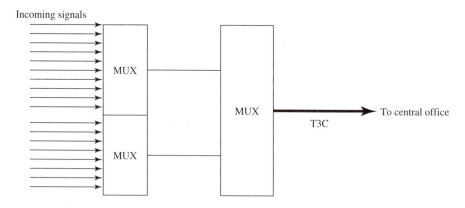

Figure 9.2 A Typical Fiber Optic System

served for testing, error coding, and monitoring functions. The error coding and other accompanying information are then compressed into the signal, boosting its operating speed to 100.8 Mb/s.

The resulting 100.8-Mb/s signal is then transmitted through a fiber optic cable containing two fibers. The signal in the fiber is the result of converting the electrical signals from the T1 line multiplxers to an optical signal operating as a digital, nonreturn-to-zero (NRZ) encoding. At the opposite end of the fiber, the signal is split into two parts; the error checking and other extraneous bits are removed, and the signal is converted back to electrical pulses and demultiplexed into various T1 lines.

9.3 COMPUTER NETWORKS

The computer network concept has grown rapidly in recent years. The **Local Area Network** (**LAN**) allows computers and related devices to be interconnected in a limited area known as a **campus**. This interconnection of computers allows information to be passed from computer to computer and allows the shared use of resources such as software, databases, and peripherals (printers, memory devices, etc.). Each computer on the network (known as a **user**) is connected to the rest of the computers using one of several possible wiring schemes (or **topologies**) and data passing and sharing techniques (or *access methods*). Central resources and a certain amount of traffic control is generally housed in a designated computer on the network known as the **fileserver**. The complexity of the access, data flow, and other operations of the network generally require a designated person to coordinate them. This person is commonly referred to as a *systems operator* or *sysop*.

Network topologies vary depending on the requirements of the users and the network. Users can be connected in a circle or **ring**. They can all be connected to a central line similar to side roads leading into a main street in a topology known as a **bus**. Finally, each user can have its own separated connection to a central spot (the fileserver or a point along a bus) for a **star** topology. The various topologies are illustrated in Figure 9.3

Different topologies have different methods of data flow. A ring network passes data from user to user around the ring. Each user removes the data that it requires and adds data to send to other users. A star network passes tailored data to each user separately, and the bus network passes data down the central line, sidetracking it to users as necessary.

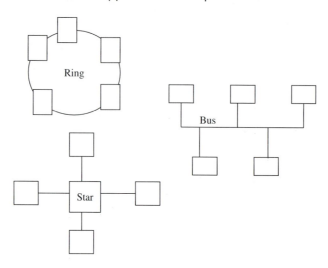

Figure 9.3 Network Topologies

Access methods for a network may involve an unlimited transmission and reception of data with provisions for conflict between users (known as **carrier sense multiple access with collision detection** or **CSMA/CD**). Data may also be contained in a package, or **token**, which is passed to each user allowing them to retrieve or send data when they have the token.

Common network packages include the **Ethernet** system which uses CSMA/CD and a bus system to connect users. Also popular is the **token ring** system and **LANtastic**. These systems also define the format of the data, the speed at which it is transmitted, and the types of signals used to represent the data.

Optical fiber may be used as the main line in a system such as a bus topology (an application known as a **fiber optic backbone**). Connections to this line then convert the signals to electronic data to be transmitted to the computers. Computer cards are also available for accepting direct fiber connections. These cards then convert the signal to electricity. Because of the short distances normally involved in a LAN, fiber attenuation is of only minimal concern whereas data-carrying capacity and speed are the critical parameters.

The Ethernet system is an example of a LAN computer system. The system can be instituted in a star topology with a main star coupler allowing interconnection among the stations and the server. As illustrated in Figure 9.4, the rays of the star consist of optical fiber which is connected to a repeater that converts the optical signal to electrical for use by the terminals on the network. Fiber used in this sort of system is

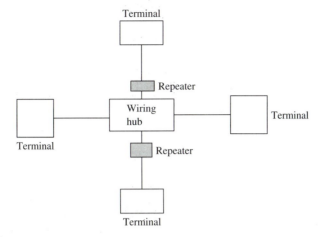

Figure 9.4 The Ethernet System

normally multimode, graded index fiber operating with an LED source and PIN detector. Since the distances involved are generally not large, larger signal powers are not necssary.

A different sort of network is now gaining popularity. This network connects computers across the country and around the world using telephone lines. To communicate on this network, computers must first convert their digital signals to analog so that they can be sent over the phone line (which is designed to accept only analog voice signals) using a **modem** (an abbreviation for modulator demodulator). The signal from these computers is then linked into larger systems using designated digital lines to connect the main systems on the network. The result is a highly flexible, ever-changing connection of systems that cover the globe. Examples of this type of network include **Bitnet** and **Internet**. These types of networks are the first stages of the Information Superhighway.

9.4 CABLE TELEVISION

Transmission of television signals to homes using copper cable is now standard practice in the United States. In the future, this copper wire will be replaced with optical fiber. The fiber will allow reception of even more TV channels as well as telephone and computer data. To understand cable television transmission with fiber, some basic concepts need to be defined.

Cable television uses **frequency division multiplexing** to transmit TV stations. Since television was originally sent via electromagnetic

(TV) waves, each channel broadcasts at a different frequency. Television sets then used a tuning system to select the desired frequency. Cable television emulates this process.

In cable TV, each station has its own electrical signal frequency. Since the cable company is transmitting the signal, they can choose which frequency to use for which station. This is why different cable systems have shows on different channels.

To transmit cable TV via fiber, signals may be used in their standard analog form to modulate the light signal at different frequencies. It is also possible to convert the TV signal to a digital format for transmission. In either case, the signals are then **time division multiplexed** (part of each signal is sent one at a time) onto the optical signal. The receiving end can then demultiplex the signal and allow the user to select which one they want to view. Digital systems predominate in fiber cable TV since the multiplexing process is greatly simplified.

9.5 FIBER OPTIC SYSTEM EXAMPLES

To illustrate the use of optical fiber in communications, the following is a step-by-step explanation of how the system works. First, there are three major factors that govern the system's properties: the physical requirements of the system such as the distance covered and the environment in which the cable will be located, the data requirements of the system, and the amount of light needed to carry the signal through the possible sources of attenuation.

The physical environment where the fiber is contained affects the type of cable used (indoor, outdoor, buried, etc.) and also how often the signal has to be repeated. For example, a local area network installation requires a consideration of plenum space, pull lengths, routing paths, and other constraints involved with installing cable in a confined space. Long-distance applications require consideration of the strength and durability of the cable, location, number of repeaters, etc. Table 9.2 provides a list of some of the physical factors to be considered.

Table 9.2 Physical Factors in Long-distance Applications

Location of transmitting, receiving, and repeating equipment
Distance between transmitter and receiver (and repeaters)
Routing plan for cable
Environment for cable (ducts, aerial, buried, etc.)
Temperature of environment
Water/moisture levels of environment
Installation constraints such as pull lengths

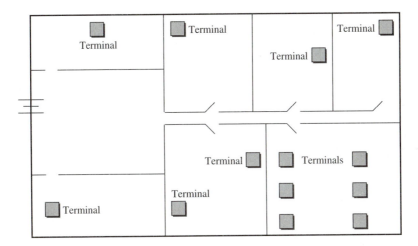

Figure 9.5 Sample Physical Environment

Figure 9.5 shows a modified blueprint of a fiber system's physical environment. A ring topology LAN is to be installed in a single building with cable pulled through conduit above a suspended ceiling. Because the LAN covers a short distance, no fiber repeaters are needed to boost signal strength. Each terminal in the LAN must be connected to the next.

Plenum-style cable can be used in this environment because the whole system is indoors. The cable installation requires runs that will allow the cable to be run freely without forcing it alongside other cable, a layout that avoids sharp bends (usually a bend radius should be no less than about six inches), and pullboxes every 200 to 300 feet or closer in any areas where several bends may be required. Figure 9.6 shows a proposed layout for this LAN using these factors as a guide.

Different communications systems will have different requirements of data capacity and operating speed. In general, the required signal-to-noise ratio and the bandwidth (or data rate) are good starting points. From the first, the minimum optical power at the receiver can be calculated; from the second, the number of channels and/or the bandwidth of the system is derived. A transmitter and receiver system can then be selected with the proper operating speeds and an accompanying fiber type with the proper capacity.

The minimum optical power at the receiver leads to a calculation of the losses in the transmission of the signal. Losses due to all connections, splices, links to transmitter/receivers, and attenuation and distortion from the fiber itself must be considered in order to arrive at a value

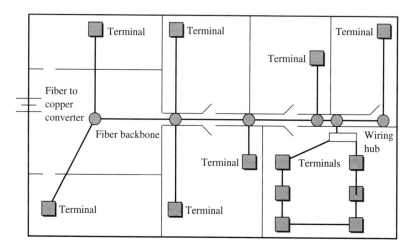

Figure 9.6 Cable Routing

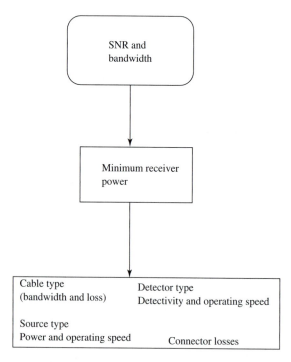

Figure 9.7 Loss Budget Procedure

for the injected power. The criteria used to calculate losses are known as the *loss* (or *power*) *budget* for the system.

To calculate the loss budget for a system, the required data rate (bit error rate), length of the transmission, minimum optical power required at the receiver, and the **dynamic range** (maximum allowed fluctuation in power) are used as starting points. Loss is calculated for the fiber-to-transmitter connection, connectors, splices, fiber loss per km, and fiber-to-receiver connection. A buffer is then added to allow for degradation of the equipment and possible increases in data requirements. The total losses are then used to determine the minimum system requirements. Figure 9.7 illustrates this loss budget procedure.

CHAPTER REVIEW

New Terms

Section 9.1
POTS   HDTV
FDDI   ISDN
Category 5   POF
CATV   POS

Section 9.2
Servicing Area Interface   Trunk Line
Central Office   Voice Channel
Prefix   DS1 or T1
Exchange   DS3 or T3
Long-distance Toll Center   T3C
Local Access Transport Area   T4

Section 9.3
Local Area Network (LAN)   Carrier Sense Multiple Access with
Campus   Collision Detection (CSMA/CD)
User   Token
Topology   Ethernet
Access Method   Token Ring
Filesaver   LANtastic
System Operator (sysop)   Fiber Optic Backbone
Ring   Modem
Bus   Bitnet
Star   Internet

Section 9.4
Frequency Division Multiplexing
Section 9.5
Loss (Power) Budget

Other Chapters with Related Information

Chapter 1
Chapter 8

Review Questions

1. Discuss some common applications of optical fiber in communications and their special attributes.
2. Draw a diagram of a typical telephone system illustrating the major connection points.
3. What aspects of cable television would capitalize on the advantages of optical fiber?
4. Sketch three common LAN topologies. Describe their features.
5. What are the transmission rates for T1, T2, and T3 lines?
6. What is LAN collision detection?
7. What is a sysop?

Thought Questions

1. Contact your local phone service provider and ask them for information about the types of systems they use in your area. What types of communications lines would you expect to find?
2. What types of communications systems would you expect to find at a college or university that acts as a main link on the Internet or similar network?
3. What kinds of services could be provided to you at home if a fiber cable were routed to your house. Would cable TV change?

REFERENCES

Cherin, Allen H. *An Introduction to Optical Fibers*. New York: McGraw-Hill, 1983.
O'Shea, Donald. *Elements of Modern Optical Design*. New York: John Wiley & Sons, 1985.

Senoir, John M. *Optical Fiber Communications—Principles and Practice.* Upper Saddle River, NJ: Prentice Hall International, 1985.

Sterling, Donald J. *Technician's Guide to Fiber Optics.* Albany, NY: Delmar Publishers, Inc., 1987.

Sze, S. M. *Semiconductor Devices—Physics and Technology.* New York: John Wiley & Sons, 1985.

Zanger, Henry, and Zanger, Cynthia. *Fiber Optics—Communications and Other Applications.* Upper Saddle River, NJ: Merrill/Prentice Hall, 1991.

10

OPTICAL FIBER MEASUREMENT AND TESTING

10.1 INTRODUCTION

A knowledge of testing and measurement procedures is essential when working with fiber optics. Tests are made in the field to determine the location and source of faults and to determine the properties of a run of fiber for future reference. In the lab, the measurements are made to check the properties of a fiber against manufacturing standards and as a method for assessing the outcome of experimental work. As in most technical fields, there are standards for testing optical fibers. Test procedures and performance standards have been established by the Electrical Industries Association (EIA), and there are independent laboratories that check commercially available fiber against these specifications.

Tests of optical fiber can be divided into *field tests* and *laboratory tests*. The first is used by installation technicians and design engineers to ensure proper performance of fiber systems; the second is used by lab and quality control technicians to verify the properties of fiber during manufacturing or before shipping. Because field tests and laboratory tests often differ in procedure, purpose, and equipment used, this chapter discusses the two types of tests separately. To help to clarify which measurements are important in a particular setting, Table 10.1 summarizes the types of tests discussed in this chapter, whether they are performed in the field or in the lab, and the section in this chapter where they are covered.

135

Table 10.1 Fiber Optic Tests

Test Name	Field/ Lab	Parameters Tested	Section
OTDR Analysis	Field	Loss/length, junction loss, and location	10.2.5
Cutback Method	Lab/ Field	Loss/length	10.2.2
Insertion Loss	Lab/ Field	Loss of fiber or connector	10.2.3
Loss Due to Scattering	Lab	Effects of Rayleigh scattering in fiber	10.3.1
Loss Due to Absorption	Lab	Amount of light absorbed by fiber	10.3.1
Dispersion	Lab	Bandwidth/change in pulse width	10.3.2
Core/Cladding Diameter	Lab	Size of the core and cladding	10.3.3
Index of Refraction Profile	Lab	Index of refraction of core	10.3.4
Numerical Aperture	Lab	NA of Fiber	10.3.5

10.2 EQUIPMENT USED IN FIELD TESTING

The equipment used in testing optical fibers in the field can be divided into two general types: the **power meter** and the **optical time domain reflectometer** (**OTDR**). Both types are based on the idea of injecting light into a fiber and then measuring the changes that the light undergoes as the result of propagation through the fiber. They both use a light source and a light detector similar to the ones discussed in Chapters 7 and 8. Light sources are usually of the LED type since the special characteristics of the laser are not necessary for basic testing. Detector types vary depending on the intended use of the testing device.

10.2.1 Optical Power Meter

The optical power meter comes in many forms, but for the testing of optical fibers, meter construction is relatively standard. The meter is designed to work with both ends of a length of fiber by injecting light into one end and receiving the light at the other end. These meters are most useful when both ends of the fiber are easily accessible (for example, fiber cable still on the spool or short lengths of fiber). Optical power meters measure the amount of light lost by transmission through the fiber and generally display the data in relative units such as decibels or dBms.

The basic power meter has a source section which is essentially an LED, a connector for coupling to the fiber, a system for maintaining constant output power, and a detection section which includes a semiconductor detector, an amplifying circuit, and a circuit for driving the display mechanism.

Figure 10.1 illustrates the basic components of a power meter. A pulse of light is injected into the fiber from the LED. A portion of the LED output is sampled and fed into a control circuit which maintains the output of the LED at a constant level. After propagating through the fiber, the light reaches the detector which produces a voltage proportional to the amount of light incident upon it. The detector's output is then either (1) amplified and compared to a voltage proportional to the original power injected into the fiber, or (2) compared to a reference voltage representing 1 mW of light power. In the first case, the difference between the two is the loss caused by the fiber (any losses caused by the connections to the fiber can be compensated for by using the

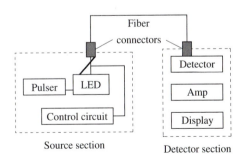

Source section   Detector section

Figure 10.1 Block Diagram of a Power Meter

techniques discussed in Sections 10.2.2 and 10.2.3) and is displayed in decibels. In the second case, the difference is displayed in dBms.

Power meters can be designed to make attenuation measurements at more than one wavelength where each wavelength requires a different source/detector combination. Sources and detectors are often packaged in easily changeable modules to facilitate wavelength changes. In the same way, meters are usually equipped with interchangeable connectors for adapting to different fibers.

10.2.2 Cutback Method

The attenuation (or loss of optical power)/length caused by the fiber can be measured using a standard technique known as the **cutback method**.

As shown in Figure 10.2, the attenuation caused by a long piece of fiber is measured using a standard power meter. The fiber is then cut back to a length of a meter or two, and the attenuation is measured again using the same launch conditions. Attenuation/length (α) is then calculated by

$$\alpha = 10 \log(P_2/P_1)/(L_1 - L_2) \qquad \textbf{(10.1)}$$

where the variables are defined as

P_1 = first power measurement

P_2 = second power measurement

L_1 = length of fiber for first measurement

L_2 = length of fiber for second measurement

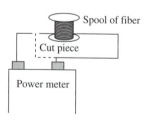

Figure 10.2 Cutback Method

10.2.3 Insertion Loss

The second method for measuring attenuation in a fiber is known as the **insertion loss** method. This technique has an advantage because it is nondestructive and doesn't require cutting the fiber. First, the power transmitted by a one-meter reference fiber (P_{ref}) is measured. Then, the reference fiber is replaced with the fiber being tested and the power transmitted (P_{test}) is measured again. The two power readings are then combined in the following formula to calculate insertion loss (IL):

$$IL = 10 \log(P_{ref}/P_{test}) \tag{10.2}$$

Since the reference fiber is a short length, the power it transmits is taken to be the input power for the test fiber, and the resulting insertion loss is actually the attenuation caused by the test fiber.

The term *insertion loss* is also used to describe the loss due to a connector or splice in a fiber. The method for measuring this loss is similar to the one just discussed. The difference is the P_{test} measurement is made by cutting the reference fiber and placing a splice or connector in it.

10.2.4 Optical Time Domain Reflectometers

The optical time domain reflectometer (OTDR) allows measurement of attenuation using only one end of the fiber. The OTDR also provides attenuation measurements along the length of the fiber in addition to an average attenuation/km. These functions, along with a graphical readout and, in some models, permanent storage of data make the OTDR a very versatile and useful instrument.

The basic structure of an OTDR is shown in Figure 10.3. Measurements are made by injecting pulses of light into one end of the fiber. As the light propagates through the fiber, a portion of it is continuously re-

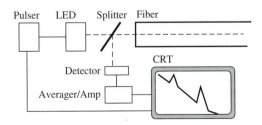

Figure 10.3 Basic Structure of an OTDR

turned to the OTDR through Rayleigh scattering (see Chapter 6). At any junction of the fiber (such as a connector, splice, break, or end of the fiber), some light is reflected to the OTDR by Fresnel reflection (see Chapter 2).

The returning light from both mechanisms is separated by the splitter and directed to the detector. The detector signal is averaged, then logarithmically amplified and used to drive the vertical deflection on the CRT. The amount of light received by the detector can be expressed as a function of both time and distance along the fiber. These two parameters are related by the velocity of light in the fiber, and either one may be used to drive the horizontal deflection of the display, although distance is usually more useful for field measurements.

The output of the OTDR is a graph (or trace) of the amount of light detected versus the position along the fiber (see Figure 10.4). The trace can be analyzed (usually with the help of circuitry in the OTDR) to determine the location of splices or connectors as well as any breaks in the fiber. The trace also provides information about the amount of attenuation caused by splices and connectors as well as the average attenuation/km of the fiber.

Since the Fresnel reflections at the junctions (connectors, splices, etc.) reflect a larger amount of light than the Rayleigh scattering of the rest of the fiber, the junctions appear as spikes in the trace. The size of the spikes will help to determine the type of junction that caused it. For example, the end of the fiber will produce a large spike because of a Fresnel reflection of about 4%, whereas a fusion splice or mechanical connection with index matching gel will reflect a good deal less and produce a much smaller spike.

10.2.5 OTDR Analysis

Attenuation calculations can be made based on the amount of power received by the OTDR. The power that is backscattered by the fiber is

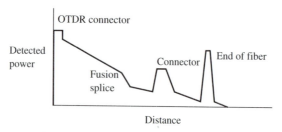

Figure 10.4 OTDR Trace

expressed as a function of time as

$$P_{\text{back}}(t) = \{[(P_0(\text{NA})^2 awV_{\text{gr}}]/8n^2)\}e^{-(aV_{\text{gr}}t)} \qquad (10.3)$$

where the variables are defined as

P_0 = input power from OTDR

NA = numerical aperture

α = scattering coefficient

w = pulse width

V_{gr} = group velocity

n = index of refraction of the core

t = time

The backscattered light must pass through the fiber twice (once as part of the original input pulse and once as it returned to the OTDR). The attenuation caused by this double pass is

$$10^{-(2a't/10 \text{ dB})} \qquad (10.4)$$

Where a' is the attenuation coefficient (dB/km) of the fiber. The power received at the OTDR is the backscattered power expressed in Equation 10.3 multiplied by the attenuation expressed in Equation 10.4

$$P_{\text{OTDR}} = P_{\text{back}}10^{-(2a't/10 \text{ dB})}$$
$$= P_0(k \ e^{-(aV_{\text{gr}}t)}) \ 10^{-(2a't/10 \text{ dB})} \qquad (10.5)$$

where $k = (\text{NA}^2 awV_{\text{gr}}/8n^2)$. Recall that attenuation is expressed in decibels by using the logarithm of the transmitted power (P_0) to the received power (P_{OTDR}). Rearranging Equation 10.5 gives an equation that can be used to determine loss as a function of time.

$$10 \log(P_{\text{OTDR}}/P_0) = 4.34(-2a'/10)[ln(k) - (aV_{\text{gr}}t)] \qquad (10.6)$$

The left side of Equation 10.6 is an expression for the attenuation of the fiber, in decibels, and the right side is the time multiplied by a constant which includes the attenuation/distance of the fiber (a'). The equation is linear, and, in simple terms, indicates that the slope of the trace shown on an OTDR corresponds to the attenuation/distance of the optical fiber tested. Using this fact, the attenuation/distance can be calculated either by hand from the data shown on the trace or electronically by the circuitry in the OTDR.

EXAMPLE 10.1

Analyze the OTDR trace shown in Figure 10.5 and determine the atten-
uation/length of the fiber, the attenuation caused by the connectors in
the fiber, the location of the connectors, and the location of the end of
the fiber.

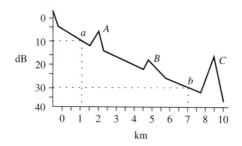

Figure 10.5 Trace for Example 10.1

To determine the attenuation/distance

Compute the slope of the trace by choosing two points on the trace (a
and b) and using the standard rise-over-run technique.

$$\text{slope} = (30 \text{ dB} - 10 \text{ dB})/(7 \text{ km} - 1 \text{ km}) = 3.33 \text{ dB/km}.$$

To determine attenuation caused by the connectors and the lo-
cation of the connectors

Two connectors are shown on the trace by the spikes labeled A and B.
Both spikes are about 5 dBs tall, and this is the attenuation they cause.
Connector A is located at about 2 km along the fiber; connector B at
about 5 km.

To determine the location of the end of the fiber

The end of the fiber is shown by the final spike, C, which is about 9 km
away.

10.3 LAB MEASUREMENT TECHNIQUES

Before attempting to make any of the following measurements, it is important to note the effects of launch conditions on the results. The conditions under which light is injected (or launched) into a fiber will affect the outcome of the measurements. Most of the methods described take this into account and compensate for it in the procedure. However, in some cases, the launch conditions must be compensated for before the measurements are undertaken. This is especially true for multimode fiber.

Recall from Chapter 5 that multimode fiber provides several pathways (or modes) through which the light travels. When light is initially injected into the fiber, it may not fill all of these modes (the *underfilled condition*), or it may fill all of the modes and spill into the cladding (the *overfilled condition*), depending on the launch conditions. As the light propagates through the fiber, it will eventually fill all of the modes, while light in the cladding will escape producing an *equilibrium condition*. Since equilibrium is the condition most likely to be found in fiber used in a communications setup, measurements on multimode fiber must be made under this condition.

Unfortunately, obtaining equilibrium naturally requires propagating light through a long piece of fiber which is not always practical when testing fiber in lab conditions. To force the light into equilibrium, two devices have been developed. The first device is known as a *mode mixer* or *mode scrambler*. It forces equilibrium by stressing the fiber and causing microbends which help to distribute the light evenly. The mode scrambler may be as simple as two heavy weights placed over sandpaper with the fiber sandwiched in between or as complex as a series of pins through which the fiber is woven in a pattern similar to a sine wave. A mode scrambler may also be a mandrel or drum around which the fiber is wrapped several times. Figure 10.6 shows some common mode scrambler configurations.

The second device used to improve launch conditions for testing is the *mode stripper* which is used to remove any light that is propagating through the cladding. The mode stripper is usually a bath of glycerin or any liquid with an index of refraction greater than the cladding. The buffer coating is removed along a length of the fiber which is then immersed in the bath. Because the index of refraction of the liquid is higher than the index of the cladding, light will pass out of the cladding and into the liquid.

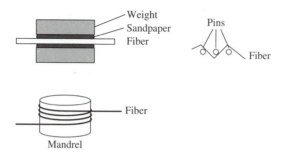

Figure 10.6 Mode Scramblers

10.3.1 Loss Due to Absorption and Scattering

For some tests, the attenuation due to a particular mechanism (instead of the total attenuation) is required. Attenuation due to absorption is often measured by detecting any increase in the temperature of the fiber. As shown by Figure 10.7(a), the fiber is placed in a tube of ground glass (to eliminate scattered light from the fiber) which is thermally connected to a temperature-measuring device such as a thermopile or a platinum resistance wire. Light absorbed by the fiber causes an increase in temperature which is detected by the temperature-measuring device. Fiber loss can then be calculated from the detected temperature.

Attenuation due to scattering can be measured by passing the fiber through a cube which is lined with six solar cells and contains index matching liquid (see Figure 10.7(b)). The light scattered by the fiber is then detected by the solar cells. The cube and solar cell arrangement is known as a Tyne's cell. The same measurement may also be made with a photodetector and integrating sphere.

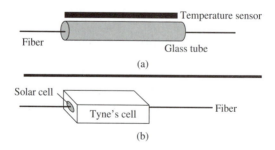

Figure 10.7 Methods for Measuring (a) Absorption and (b) Scattering

10.3.2 Dispersion

Dispersion loss measurements can be made in terms of pulse width broadening or in terms of bandwidth. Recall from Chapter 6 that the amount that a pulse broadens as it passes through a fiber is commonly known as *dispersion* and is normally used to describe the information-carrying capacity of a single-mode fiber. For multimode fiber, however, the information-carrying capacity is often expressed as *bandwidth* or *bandwidth length product* which is proportional to the maximum frequency that a fiber can transmit through a certain distance. The two properties are mathematically related (see Chapter 6). Measurements of pulse width are known as *time domain measurements* since the results are times corresponding to how much the pulse width has increased. Bandwidth measurements are known as *frequency domain measurements* since the results are frequencies.

To measure dispersion (pulse width), a pulse of light is sent through an optical fiber and detected at the opposite end. The width of the transmitted pulse is compared to the width of the original to determine the dispersion. The basic setup is illustrated in Figure 10.8. Pulse widths may be measured at half of the pulse height (or 3 dBs) or the rms pulse widths may be used. The change in pulse width divided by the fiber length corresponds to the dispersion of the fiber. The bandwidth of the fiber can be calculated from the dispersion if we assume the pulse is Gaussian. Bandwidth (BW) length product is related to the dispersion (*D*) by

$$\text{BW} = 0.44/D \tag{10.7}$$

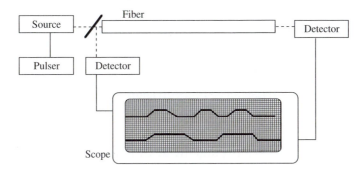

Figure 10.8 Pulse Width Measurement

EXAMPLE 10.2

The scope display shown in Figure 10.9 is from a dispersion measurement setup. If the fiber is 10 km long, use the data from the scope to determine the dispersion of the fiber. Calculate pulse width at the 3-dB points. Use your result to determine bandwidth length product.

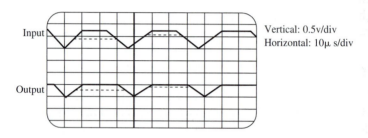

Figure 10.9 Trace for Example 10.2

To determine pulse width at the 3-dB points

Using the one pulse for each trace (first pulse on the diagram), determine the peak voltage.

3 divisions × 0.5 volts/division = 1.5 volts (input trace)

1 division × 0.5 volts/division = 0.5 volts (output trace)

Divide the peak voltage by two and measure the width of the pulse at this point.

1.5 volts/2 = 0.75 volts (input trace)

pulse width = 2.3 divisions × 10 μs/division = 23 μs (input trace)

0.5 volts/2 = 0.25 volts (output trace)

pulse width = 3 divisions × 10 μs/division = 30 μs (output trace)

To determine dispersion

Subtract the two pulse widths and divide by the length of the fiber.

3-dB dispersion = (30 μs − 23 μs)/10 km = 0.7 μs/km

To calculate bandwidth length product

Use Equation 10.7 where the dispersion is given above.

$$BW = 0.44/0.7 \ \mu s/km = 0.63 \ MHz\text{-}km$$

Bandwidth can also be measured directly by sending a modulated beam of light through the fiber and detecting its output with a spectrum analyzer.

10.3.3 Core/Cladding Diameter

The diameter of the core and the cladding of a fiber can be determined experimentally. The simplest method is shown in Figure 10.10. Visible light is injected into a short piece of fiber, and the exiting light is projected on a screen. The core and cladding will be easily discernible for a step index fiber, and the ratio between the two can be measured by measuring the image. Cladding diameter can then be measured using a micrometer, and the core diameter can be calculated from the ratio.

10.3.4 Index of Refraction Profile

To measure the index of refraction profile of a fiber, a slice (or slab) of the fiber may be cut off and polished flat on both sides. The slab can then be placed in the test arm of a Michelson interference microscope as shown in Figure 10.11. The light in the microscope is passed through a beam splitter which divides it into the reference arm and the test arm. Light in the test arm passes through the slab of optical fiber, reflects

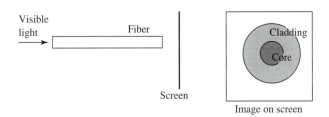

Figure 10.10 Core/Cladding Diameter Measurement

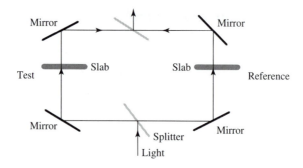

Figure 10.11 Interference Microscope

from the mirror, and recombines with the reference arm. Light in the reference arm passes through a reference slab with a constant index of refraction. The varying indices of refraction in the test slab will cause phase shifts that show up as interference fringes when the reference arm and the test arm are recombined. The pattern produced can be photographed or detected by a CCD camera and then analyzed to determine the index of refraction. The shifted fringes shown in the image of Figure 10.12 are compared to the straight fringes. By counting the number of fringes that the shifted fringes are displaced (N), the difference between the index of refraction of the test slab, and the reference slab (Δn) can be calculated by

$$\Delta n = (N\lambda)/x \qquad (10.8)$$

where λ is the wavelength used and x is the thickness of the slabs.

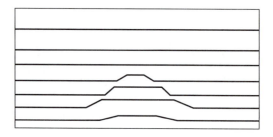

Figure 10.12 Image from Microscope

Figure 10.13 Near-Field Scanning Method

Index of refraction profile can also be determined by the refractive near-field method shown in Figure 10.13. Light from an LED is injected into the fiber and the light escaping from the opposite end is detected by the photodiode. The photodiode moves across the face of the fiber, and power versus position is recorded. The curve formed by these data is proportional to the index of refraction profile although a correction factor must be applied to compensate for mode effects.

10.3.5 Numerical Aperture

The numerical aperture of step index optical fibers can be measured using the far-field output pattern of the fiber. As shown in Figure 10.14, the fiber is mounted on a rotary stage, and a photodetector is fixed at the output end. Readings from the detector are made as the fiber is rotated. The readings are plotted to form a curve similar to the one shown in the inset of Figure 10.14. Numerical aperture is then determined by measuring the width of the curve (θ) at 10% of the peak power and using

$$NA = \sin(\theta/2) \tag{10.9}$$

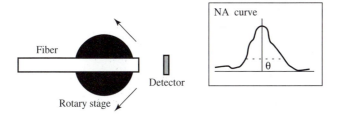

Figure 10.14 Measurement of Numerical Aperture

EXAMPLE 10.3

The following data was taken from a numerical aperture measurement. Determine the numerical aperture of the fiber.

Angle (°)	Detector Output (V)
30	0.001
25	0.98
20	2.01
15	2.96
10	4.04
5	4.99
0	6.01
−5	5.00
−10	4.01
−15	2.99
−20	2.00
−25	1.00
−30	0

Plot the data given in the table.

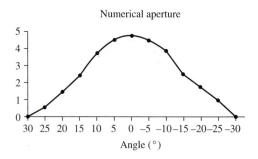

Determine the 10% of peak points on the graph by multiplying the maximum detector reading (5) by 10%.

$$5 \times 10\% = 0.5 \text{ V}$$

Measure the width of the plot at 0.5 volts.

$$\text{Width} = 50°$$

Use Equation 10.9 to determine the numerical aperture.

$$\text{NA} = \sin(50/2) = 0.423$$

CHAPTER REVIEW

New Terms

Section 10.1
    Field Tests
    Lab Tests

Section 10.2
    Cutback Method
    Insertion Loss
    Optical Time Domain Reflectometer
    Power Meter

Section 10.3
    Time Domain
    Frequency Domain

Other Chapters with Related Information

    Chapter 2
    Chapter 5
    Chapter 6
    Chapter 7
    Chapter 8

Review Questions

1. Give two examples of field tests and two examples of lab tests.
2. Why are mode scramblers used?
3. What causes light to return along the fiber to the OTDR?
4. What are the advantages of an OTDR over a power meter?
5. What is a Tyne's cell?

6. Describe how the index of refraction profile of a fiber can be measured with a photodetector.

REFERENCES

Cherin, Allen H. *An Introduction to Optical Fibers*. New York: McGraw-Hill, 1983.

Davenport, Craig. "Fiber Optics Standards That Demand Attention: Performance, Material Quality and Safety," *Photonics Spectra,* May 1991.

Hentshel, Christian. *Fiber Optics Handbook*. Hewlett-Packard, 1984.

Lidh, John S. "Choose the Right OTDR for Troubleshooting Your System," *Cabling Installation Maintenance,* April/May 1993.

Senoir, John M. *Optical Fiber Communications—Principles and Practice*. Upper Saddle River, NJ: Prentice Hall International, 1985.

Sterling, Donald J. *Technician's Guide to Fiber Optics*. Albany, NY: Delmar Publishers Inc., 1987.

Zanger, Henry, and Zanger, Cynthia. *Fiber Optics—Communications and Other Applications,* Upper Saddle River, NJ: Merrill/Prentice Hall, 1991.

✳ APPENDIX A: SAFETY

Two kinds of hazards can occur when working with optical fiber: glass shards and optical radiation. Aside from other common laboratory hazards such as electrical shock and chemical exposure, these two hazards are most prominent. The use of a glass optical fiber means that the possibility of glass shards or slivers is always strong, especially when fibers are cut, cleaved, or broken. These slivers are dangerous because they are often difficult to see and can easily embed themselves into the skin so that the potential for cuts, contamination of the eyes, or even swallowing is increased.

To reduce the hazard of glass shards, care should be taken when handling the fiber, especially if it is not part of a cable package. Always wear safety goggles to protect your eyes and carefully wash your hands after working in the lab or the field.

Optical radiation hazards (especially from lasers) can cause damage to the eyes and is especially dangerous when working with invisible infrared light used with fiber. The Center for Disease and Radiological Health (CDRH) and the American National Standards Institute have established guidelines and regulations for working with lasers and other optical hazards. These guidelines stipulate a labeling system that warns the user of possible dangers and necessary precautions. Using specialized protective goggles, shutters, and other systems can limit or deny exposure to hazardous emissions.

APPENDIX B: USEFUL CONSTANTS AND DATA

General Physical Constants

Permeability in vacuum	μ_0	1.25663×10^{-8} H/c
Permittivity in vacuum	ε_0	8.85418×10^{-14} F/cm
Planck's constant	h	6.62617×10^{-34} J-s
Speed of light in vacuum	c	2.99792×10^8 m/s

Bandgap Energies (in eV) of Common Semiconductors at 300 K

Ge	0.66
Si	1.12
SiC	2.99
AlSb	1.58
GaAs	1.42
GaP	2.26
GaSb	0.72
InAs	0.36
InP	1.35
InSb	0.17
CdS	2.42
CdTe	1.56
ZnO	3.35
ZnS	3.68
PbS	0.41
PbTe	0.31

Group II, III, IV, and V Materials
Used in Semiconductor Devices

Group II	Group III	Group IV	Group V
Magnesium (Mg)	Boron (B)	Carbon (C)	Nitrogen (N)
Zinc (Zn)	Aluminum (Al)	Silicon (Si)	Phosphorous (P)
Cadmium (Cd)	Gallium (Ga)	Germanium (Ge)	Arsenic (As)
Mercury (Hg)	Indium (In)	Tin (Sn)	Antimony (Sb)
		Lead (Pb)	

Index of Refraction of Common Optical Materials

Material	at 656.3 nm	at 589.2 nm	at 486.1 nm
Barium Flint	1.588	1.591	1.598
Borosilicate Crown, 1	1.498	1.500	1.505
Borosilicate Crown, 2	1.515	1.517	1.523
Borosilicate Crown, 3	1.509	1.511	1.517
Light Flint, 1	1.571	1.575	1.585
Light Flint, 2	1.520	1.576	1.586
Dense Flint, 2	1.612	1.617	1.629
Dense Flint, 4	1.644	1.649	1.663
Extra Dense Flint	1.456	1.458	1.463
Light Barium Crown	1.538	1.541	1.547
Quartz	1.456	1.458	1.463

GLOSSARY

absorption the loss of light power caused by material in the fiber (especially moisture) absorbing some of the light.

acceptance angle the area within which light will enter the fiber at the proper angle and propagate through the core.

A/D conversion of analog communication signals to digital communication signals.

amplitude the height of a wave.

amplitude modulation (AM) an analog communication method that alters the amplitude of the carrier signal in order to represent data.

analog smooth and continuous values such as waves.

angular misalignment the misalignment of connected fibers caused by the two fiber ends being something other than parallel to each other.

asynchronous A form of communication in which the transmitter and receiver operate on separate time signals.

attenuation loss of optical power.

avalanche photodiode (APD) an optical detector based on the *p-n* junction but using an increased electric field to cause multiple electrons to be produced by a single incident photon.

bait rod a glass rod used in the manufacture of optical fiber.

band gap energy the amount of energy required to move an electron from the valence band to the conduction band.

bandwidth the range of frequencies that a communication system can transmit with acceptable loss (less than 3 dBs).

bandwidth length product (BWL) a measurement of dispersion loss in a fiber.

biphase code a digital encoding system that uses transitions in the optical power to represent bits.

bit one binary number.

Bitnet one of many large-scale networks of computers.

buffer protective layer over fiber cladding.

bus a computer network topology.

byte eight bits.

campus the geographical area encompassed by a LAN.

carrier sense multiple access with collision detection (CSMA/CD) a method of transmitting data in a LAN.

carrier signal the signal that carries information in a communications system (for example, radio waves).

Category 5 a standard for fiber optic test equipment.

CATV cable television.

cladding the outer layer of an optical fiber.

cleaving the breaking of a solid along natural lines such as along the scratch on the surface.

coherence a set or constant phase between light waves.

communication the process of conveying information from one place to another.

conduction band the energy levels at which electrons are freed from an atom.

connector a removable joint between two fibers.

constructive interference the combination of in-phase waves to produce a larger wave that has an amplitude equal to the sum of the individual wave.

core the inner rod of an optical fiber.

coupler a connection that allows light to be separated into more than one fiber.

covalent bond a bond between atoms formed by sharing an electron.

critical angle the angle at which internal reflection occurs.

cutback method a method for measuring the loss/length of a fiber.

cutoff frequency the signal frequency at which the output of a photodetector drops by 3 dB.

cutoff wavelength the wavelength below which a detector cannot produce a useful signal.

cyclic redundancy check (CRC) an error-checking scheme used in communications.

D/A digital-to-analog conversion.

dark current current produced by a detector which is not caused by incident light.

decibel (dB) a logarithmic loss unit.

dee-star (D*) a measure of detector sensitivity.

demultiplexing the separation of signals that are sent through a fiber simultaneously.

depletion region area in a *p-n* junction where holes and electrons combine and form a neutral material through which current does not normally pass.

detectivity a measure of detector sensitivity.

dichroic material a material that can be used to cause reflection at set wavelengths through interference.

diffraction the spreading out of light caused by its passage through an opening.

diffraction grating a device that separates light by wavelength through the use of slits and diffraction.

diffuse reflection reflection from rough, opaque surfaces like wood.

digital discrete, discontinuous quantities.

dispersion the distortion of a signal in an optical fiber.

dispersion shifted fiber fiber designed to have a minimum material dispersion at 1550 nm.

doping the addition of other materials to a crystal to produce a certain semiconductor type.

dry no polish (DNP) a fiber connector.

duplex a two-fiber cable (or two-way communications).

dynamic range the maximum allowable power fluctuation in a communication system.

dynamometer a device that measures the amount of tensile force applied to a fiber cable during installation.

elastomeric splice a fiber splice that uses an elastomer (rubbery material) to hold the fibers in place.

electromagnetic spectrum an ordered list of types of electromagnetic waves. List is usually given in order of increasing frequency or wavelength.

encoding imposing data onto a carrier signal.

end separation loss in a fiber connection caused by gaps between the ends of the two fibers.

error detection a system used to identify errors in a transmitted signal.

Ethernet one of many computer networks.

excited state condition of an atom that has absorbed excess energy.

extrinsic loss loss in a fiber connection that is due to the connector itself.

fall time the time it takes an optical source to drop from 90% to 10% of its peak power.

Fiber Distributive Data Interface (FDDI) a standard for fiber communications.

fiber optic backbone a main line of optical fiber in a LAN.

fileserver a computer on a LAN that coordinates and monitors data flow.

forward biased connection of a DC voltage with the plus side to the p and the minus side to the n of a p-n junction.

four-rod splice a fiber splice that uses four rods to hold the fibers in place.

Fraunhofer diffraction diffraction that occurs when the light source and observation point are at infinity.

frequency the number of cycles/second of a wave.

frequency division multiplexing a method of sending multiple signals simultaneously by using different frequencies for each.

frequency modulation (FM) an analog system of communications that modifies the frequency of a carrier signal to represent data.

Fresnel diffraction diffraction that occurs when the light source and observation point are near the opening.

Fresnel reflection reflection from a smooth, transparent surface.

full width at half maximum (FWHM) a measurement of the width of a curve at the points which represent half the maximum.

fusion splice a splice between fibers that melts the ends together.

geometrical optics the study of light as a straight line beam.

graded index fiber fiber that uses a gradual (graded) change in index from the core to the cladding to minimize dispersion.

GRIN lens graded index lens that focuses light through changes in the index of refraction rather than the curvature of glass.

ground state an atom's condition when it has the normal (or minimum) amount of energy.

Group III and Group V materials located in a certain location in the periodic table.

group velocity velocity of all the electromagnetic waves traveling together.

heterojunction a p-n junction made of different materials.

high-definition television (HDTV) a new system of television communications that promises better picture quality and more sound capabilities.

hole a gap that an electron could fill.

homojunction a p-n junction made of the same materials.

hybrid a fiber cable that also includes copper wire.

index matching gel a liquid used to minimize Fresnel reflection in connectors by replacing the air in any gaps.

index of refraction a property of a material that indicates the speed at which light travels through it.

index of refraction profile a plot of the index of refraction in a cross section of a fiber.

infrared the area of the electromagnetic spectrum that consists of waves with slightly longer wavelengths than visible light.

insertion loss a method for determining the amount of loss caused by a connector or splice. Also, a method for determining loss in a fiber.

Integrated Services Digital Network (ISDN) a communications standard for fiber.

intelligence signal the data to be sent in a communications system.

interference an effect observed in the combination of coherent light with different phases.

Internet one of many large-scale computer networks.

intrinsic layer a semiconductor material in a p-n junction that is neither p- nor n-type.

intrinsic loss loss in a fiber connection due to the fiber.

irradiance the power of light incident on a detector surface divided by the area of the detector.

jacket the outer protective layer in a fiber cable.

Johnson noise see *Thermal noise*.

LANtastic a type of LAN design.

laser light amplification by stimulated emission of radiation. A device that produces monochromatic, coherent, directional light at high powers through a special structure.

lateral misalignment misalignment by a lateral shift of two connected fibers.

light-emitting diode (LED) a semiconductor device based on the *p-n* junction but designed to produce light.

local area network (LAN) a group of computers linked together and contained within a relatively small geographical region.

loss a reduction in the amount of light or a distortion of the signal that it is carrying.

Manchester code a digital encoding system.

material dispersion dispersion caused by the fact that different wavelengths travel at different speeds.

meridional ray a ray that passes through the optic axis.

microbends small variations on the surface of a fiber core.

microlens a small lens fitted onto a detector or optical source.

Miller code a method of digital encoding.

minimum bend radius the minimum radius that a fiber cable can be bent without damaging the fiber.

minimum detectable power the minimum amount of light that a photodetector can detect.

mixing block a device that disperses light into several fibers.

modal dispersion loss in a fiber caused by light traveling through modes of different lengths.

mode a path for light to follow in a fiber.

mode distribution the distribution of light within the fiber modes.

mode equilibrium light filling all the possible modes in a fiber and no light in the cladding.

modem modulator/demodulator that is used to convert digital computer signals to analog telephone signals and vice versa.

modulation the changing of the properties of a carrier in order to represent data.

monitor photodiode used in a semiconductor laser to regulate output.

monochromatic a single wavelength or color.

multiplexing the combination of several signals onto one transmission line.

n-type a semiconductor material with an excess of electrons.

nibble four bits.

noise current in a detector that is caused by something other than incident light.

noise equivalent power (NEP) a method of rating a detector's response.

nonreturn to zero (NRZ) a method of digital encoding.

nonreturn to zero inverted (NRZI) a method of digital encoding.

numerical aperture the number value used to describe the angular limit on light as it enters or exits a fiber.

Nyquist noise see *Thermal noise*.

optical time domain reflectometer (OTDR) an instrument used to measure loss and locate breaks in a fiber.

orbital the path that an electron follows in an atom.

p-n junction a semiconductor device formed by joining a *p*-type and an *n*-type material together. (Also known as a diode.)

***p*-type** a semiconductor material with an excess of holes.

parity an error detection method used in communications. Also, the number of 1s in a digital word.

phase the alignment of one wave to another.

phase velocity velocity of a single point of constant phase on a wave.

photoconductive a device that changes conductivity when exposed to light.

photogeneration the production of electrons by incoming photons on a detector's surface.

photomultiplication the increase in electrons produced in a detector due to collisions with other electrons.

photon a particle of light.

photovoltaic a device that produces a voltage when exposed to light.

physical optics the study of light as a wave.

pigtail a section of fiber attached to a detector or source to enhance its connection to longer fiber lengths.

PIN photodiode a detector that uses an intrinsic layer to increase its area of sensitivity.

Planck's constant 6.6×10^{-34} j-s

plastic clad silica (PCS) fiber with a glass core and a plastic cladding.

plastic optical fiber (POF) fiber constructed from plastic rather than the glass normally used in communications grade fiber.

plenum a confined space (such as under the floor or between the walls) inside of a building that is used for routing cable.

polarization the orientation of a wave.

population inversion when the number of atoms that are excited exceed the number in the ground state.

power meter a device for measuring loss and output power in a fiber.

preform the glass rod used in the manufacture of optical fiber.

pulse coded modulation (PCM) a method of encoding a digital signal onto a carrier.

quantum efficiency the number of electrons produced in a detector by each incoming photon.

quantum well a semiconductor laser design that incorporates layers of material so thin that the effects of quantum mechanics become a factor in the operation.

random polarization light waves have a random orientation to each other.

Rayleigh scattering the dispersion of light caused by flaws or foreign particles in a fiber.

ray tracing the method of tracing light through a series of optical elements using the principles of refraction and Snell's Law.

recombination electrons fill holes in a p-type semiconductor material.

rectilinear propagation light traveling in a straight line.

refraction the changing in the direction of light as it passes from one material to another.

responsivity the sensitivity of a detector to incoming light.

return to zero a method of digital encoding.

reverse biased connection of a DC voltage to a p-n junction with the negative side connected to the p-type materials and the positive side connected to the n-type.

ring a computer network topology.

rise time the amount of time it takes to increase from 10% to 90% of peak power.

sample a portion of analog signal selected to convert it to digital.

scattering the redirection of light caused by flaws and cracks in a fiber.

scribing tool a device used to scratch a fiber's surface prior to cleaving.

semiconductor laser a source of light that is produced by a light amplifier constucted from a p-n junction.

sensitivity a measure of the response of a photodetector to incoming light.

shot noise noise in a detector caused by the quantum nature of photons

signal-to-noise ratio (SNR) the ratio of the strength of a signal to that of the noise in a detector.

simplex a cable with a single fiber (or one-way communication).

single mode a fiber with only one mode for the light to travel through.

skew rays rays that do not pass through the center of a fiber core.

Snell's Law relates index of refraction and incoming and outgoing angles for light passing from one material to another.

soot the stream of heated materials applied to the preform in the manufacture of optical fiber.

spatially coherent a group of waves with a fixed phase.

spectral width a range of wavelengths.

specular reflection reflection from smooth, highly reflective surfaces like mirrors.

splice a permanent joint between two fibers.

spontaneous emission the emission of light by an atom without outside stimulus.

star a computer network topology.

star coupler a device used to equally divide the light from one fiber into several other fibers.

step index fiber a fiber that has an abrupt change in index of refraction between the core and the cladding.

stimulated emission the emission of light by an atom due to outside stimulus.

strength member the component of a fiber cable that provides tensile and shear strength.

synchronous a method of communications in which both the transmitter and the receiver use the same timing signal.

temporally coherent a group of waves with the same wavelength.

thermal noise noise in detector caused by random motion of the atoms. (also known as *Johnson* or *Nyquist* noise.)

thermoelectric cooler (peltier element) a device that cools a semiconductor laser using the flow of electrons.

threshold current the current above which a semiconductor laser produces a beam.

time division multiplexing a method of transmitting several signals by splitting them into parts and transferring each part one at a time.

token a set of data that is passed from computer to computer in a token ring.

token ring a LAN design.

topology the physical and information flow layout of a LAN.

trunk line the main phone line for incoming and outgoing long-distance calls.

user computer (or person) connected to a LAN.

valence band the energy level at which electrons are still attached to an atom.

valence electron the electron farthest away from the nucleus of an atom.

valence orbital the outermost orbital of an atom.

voice channel a method of measuring the capacity of a telephone system.

waveguide dispersion loss in a fiber caused by light traveling in the cladding of a single-mode fiber.

wavelength the peak-to-peak length of a wave.

wavelength division multiplexer (WDM) a method of sending multiple signals simultaneously by using a different wavelength for each one.

word a group of bits.

zero dispersion wavelength the wavelength at which the difference in speeds between wavelengths is minimal.

INDEX